Parallel Finite-Difference Time-Domain Method

For a complete listing of recent titles
in the *Artech House Electromagnetic Analysis Series*,
turn to the back of this book.

Parallel Finite-Difference Time-Domain Method

Wenhua Yu
Raj Mittra
Tao Su
Yongjun Liu
Xiaoling Yang

ARTECH HOUSE
BOSTON | LONDON
artechhouse.com

Library of Congress Cataloging-in-Publication Data
Parallel finite-difference time-domain method/Wenhua Yu ... [et al.].
  p. cm.—(Artech House electromagnetic analysis series)
  ISBN 1-59693-085-3 (alk. paper)
  1. Electromagnetism—Data processing. 2. Parallel programming (Computer science).
  3. Finite differences—Data processing. 4. Time-domain analysis. I. Yu, Wenhua. II. Series.

QC760.54.P37 2006
537—dc22   2006045967

British Library Cataloguing in Publication Data
Parallel finite-difference time-domain method.—(Artech House electromagnetic analysis series)
  1. Electromagnetism—Computer simulation 2. Finite differences 3. Time-domain analysis
  I. Yu, Wenhua
  537'.0113
  ISBN-10: 1-59693-085-3

Cover design by Igor Valdman

© 2006 ARTECH HOUSE, INC.
685 Canton Street
Norwood, MA 02062

All rights reserved. Printed and bound in the United States of America. No part of this book may be reproduced or utilized in any form or by any means, electronic or mechanical, including photocopying, recording, or by any information storage and retrieval system, without permission in writing from the publisher.

All terms mentioned in this book that are known to be trademarks or service marks have been appropriately capitalized. Artech House cannot attest to the accuracy of this information. Use of a term in this book should not be regarded as affecting the validity of any trademark or service mark.

International Standard Book Number: 1-59693-085-3
ISBN-10: 1-59693-085-3
ISBN-13: 978-1-59693-085-8

10 9 8 7 6 5 4 3 2 1

Contents

Preface ix

Chapter 1 FDTD Method 1
    1.1 FINITE-DIFFERENCE CONCEPT 1
    1.2 INTRODUCTION TO FDTD 3
    1.3 NUMERICAL DISPERSION 8
    1.4 STABILITY ANALYSIS 10
    1.5 NONUNIFORM MESH 11
    REFERENCES 13

Chapter 2 Boundary Conditions 15
    2.1 PEC AND PMC BOUNDARY CONDITIONS 16
    2.2 MUR'S ABSORBING BOUNDARY CONDITION 18
    2.3 UNSPLIT PML 21
    2.4 STRETCHED COORDINATE PML 26
    2.5 TIME CONVOLUTION PML 27
    REFERENCES 32

Chapter 3 Improvement of the FDTD 33
    3.1 CONFORMAL FDTD FOR PEC OBJECTS 33
    3.2 CONFORMAL TECHNIQUE FOR DIELECTRIC 40
    3.3 ALTERNATING DIRECTION IMPLICIT (ADI) ALGORITHM 42
        3.3.1 Update for ADI Equations 43
        3.3.2 PML Truncation Techniques for ADI-FDTD 47
        3.3.3 Separable Backward-Central Difference 49
    3.4 SIMULATION OF DISPERSIVE MEDIA 53
        3.4.1 Recursive Convolution 53
        3.4.2 Pole Extraction Techniques 58
        3.4.3 Simulation of Double-Negative Materials 59

3.5 CIRCUIT ELEMENTS 60
REFERENCES 67

Chapter 4 Excitation Source 69

4.1 INTRODUCTION 69
4.2 TIME SIGNATURE 71
4.3 LOCAL SOURCE 73
4.4 SOURCES FOR UNIFORM TRANSMISSION LINES 78
4.5 POWER INTRODUCED BY LOCAL EXCITATION SOURCES 87
4.6 PHASE DIFFERENCE AND TIME DELAY 89
4.7 PLANE WAVE SOURCES 90
REFERENCES 95

Chapter 5 Data Collection and Post-Processing 97

5.1 TIME-FREQUENCY TRANSFORMATION 97
5.2 MODE EXTRACTION OF TIME SIGNAL 100
5.3 CIRCUIT PARAMETERS 103
5.4 NEAR-TO-FAR-FIELD TRANSFORMATION 106
    5.4.1 Basic Approach 106
    5.4.2 Far-Field Calculation for Planar Structures 109
    5.4.3 Data Compression and Its Application 113
REFERENCES 117

Chapter 6 Introduction to Parallel Computing Systems 119

6.1 ARCHITECTURE OF THE PARALLEL SYSTEM 120
    6.1.1 Symmetric Multiprocessor 121
    6.1.2 Distributed Shared Memory System 122
    6.1.3 Massively Parallel Processing 123
    6.1.4 Beowulf PC Cluster 124
6.2 PARALLEL PROGRAMMING TECHNIQUES 125
6.3 MPICH ARCHITECTURE 126
6.4 PARALLEL CODE STRUCTURE 130
6.5 EFFICIENCY ANALYSIS OF PARALLEL FDTD 135
REFERENCES 143

Chapter 7 Parallel FDTD Method 145

7.1 INTRODUCTION TO THE MPI LIBRARY 146

7.2 DATA EXCHANGING TECHNIQUES 147
7.3 DOMAIN DECOMPOSITION TECHNIQUE 150
7.4 IMPLEMENTATION OF THE PARALLEL FDTD METHOD 152
    7.4.1 Data Exchange Along the x-Direction 156
    7.4.2 Data Exchange Along the y-Direction 162
    7.4.3 Data Exchange Along the z-Direction 163
7.5 RESULT COLLECTION 165
    7.5.1 Nonuniform Mesh Collection 165
    7.5.2 Result Collection 166
    7.5.3 Far-Field Collection 167
    7.5.4 Surface Current Collection 167
7.6 ASSOCIATED PARALLEL TECHNIQUES 170
    7.6.1 Excitation Source 170
    7.6.2 Waveguide Matched Load 173
    7.6.3 Subgridding Technique 173
7.7 NUMERICAL EXAMPLES 174
    7.7.1 Crossed Dipole 175
    7.7.2 Circular Horn Antenna 175
    7.7.3 Patch Antenna Array Beam Scanning 176
REFERENCES 178

Chapter 8 Illustrative Engineering Applications 179

8.1 FINITE PATCH ANTENNA ARRAY 179
8.2 FINITE CROSSED DIPOLE ANTENNA ARRAY 185
REFERENCES 192

Chapter 9 FDTD Analysis of Bodies of Revolution 193

9.1 BOR/FDTD 193
9.2 UPDATE EQUATIONS FOR BOR/FDTD 194
9.3 PML FOR BOR/FDTD 198
9.4 NEAR-TO-FAR-FIELD TRANSFORMATION IN BOR/FDTD .. 200
9.5 SINGULAR BOUNDARY IN THE BOR/FDTD 207
9.6 PARTIALLY SYMMETRIC PROBLEM 208
    9.6.1 Normally Incident Plane Wave 209
    9.6.2 Obliquely Incident Plane Wave 210
REFERENCES 213

Chapter 10 Parallel BOR/FDTD 215

10.1 INTRODUCTION TO PARALLEL BOR/FDTD 215
10.2 IMPLEMENTATION OF PARALLEL BOR/FDTD 216
10.3 EFFICIENCY ANALYSIS OF PARALLEL BOR/FDTD 220
10.4 REFLECTOR ANTENNA SYSTEM 222
   10.4.1 Reflector Antenna Simulation 222
   10.4.2 Combination with Reciprocity Principle 224
   10.4.3 Simulation of the Reflector Antenna System 226
REFERENCES 228

Appendix A Introduction to Basic MPI Functions 229

A.1 SELECTED MPICH FUNCTIONS IN FORTRAN 229
A.2 SELECTED MPICH FUNCTIONS IN C/C++ 236
A.3 MPI DATA TYPE 241
A.4 MPI OPERATOR 242
A.5 MPICH SETUP 243
   A.5.1 MPICH Installation and Configuration for Windows 243
   A.5.2 MPICH Installation and Configuration for Linux 243
REFERENCES 244

Appendix B PC Cluster-Building Techniques 245

B.1 SETUP OF A PC CLUSTER FOR THE LINUX SYSTEM 245
   B.1.1 PC Cluster System Description 246
   B.1.2 Linux Installation 246
   B.1.3 Linux System Configuration 248
   B.1.4 Developing Tools 251
   B.1.5 MPICH 252
   B.1.6 Batch Processing Systems 253
B.2 PARALLEL COMPUTING SYSTEM BASED ON WINDOWS . 253
   B.2.1 System Based on Windows 254
   B.2.2 System Based on a Peer Network 254
B.3 USING AN EXISTING BEOWULF PC CLUSTER 254
REFERENCES 256

List of Notations 257

About the Authors 259

Index 261

Preface

As an important branch of electromagnetic fields and microwave techniques, computational electromagnetics has found broad applications in scientific research and engineering applications. Commonly used methods in computational electromagnetics include the finite element method (FEM), the finite difference time domain (FDTD) method, and the method of moments (MoM), and they all find applications to the solution of a wide variety of electromagnetic problems.

The FDTD method is well suited for analyzing problems with complex geometrical features as well as those containing arbitrarily inhomogeneous materials. Also, the FDTD method does not require the derivation of a Green's function or the solution of a matrix equation that is needed in both the MoM and FEM. However, the FDTD technique does place a relatively heavy burden on computer resources when it is used to analyze a complex problem that occupies a large computational volume or includes very fine features.

A single processor with 2 gigabytes of memory is able to handle a problem whose dimensions are on the order of 10 wavelengths. Often the problem size that we need to tackle is far beyond the capability of a single processor in both terms of simulation time and physical memory. One approach to circumventing this problem is to resort to a parallel processing technique that employs multiple processors to work concurrently on a single task, with the processors sharing the burden both in terms of memory use and computational effort. Each processor in a parallel computing system only handles a part of the entire problem, and exchanges information between each other via a communication protocol. An important measure by which a parallel processing method is evaluated is its efficiency. The FDTD algorithm is embarrassingly parallel in nature, and it only requires an exchange of the field among the near neighbors; this makes the FDTD algorithm very desirable for parallel processing.

Though parallel computing techniques have a long history, most of them are tailored for special computer architectures; consequently, their portability between different computers is relatively poor. However, an exception to these is the parallel algorithm that utilizes the message passing interface (MPI), which is a process-oriented standard library that utilizes a communicator, and hence, renders it highly portable.

The focus of this book, which is somewhat of a sequel to an earlier text on the FDTD published by the authors, is on parallel implementation of the FDTD method. To provide some necessary background on the FDTD method, we often refer the reader to the existing literature where details in the background material can be found. However, we do include some code segments in both the C and the Fortran programming languages that are designed to help the reader understand the structure of parallel code and the programming method on which the code is based.

The presentation in this book is divided into two parts. In the first part, Chapters 1–8, we introduce the basic concepts of the 3-D Cartesian FDTD method such as the boundary conditions, near-to-far-field transformation, and enhancements to the FDTD, and we follow this with a discussion of the parallel implementation of the FDTD method. In the second part, Chapters 9–10, we introduce the basic concepts of the body of revolution (BOR) FDTD method and discuss the topic of absorbing boundary condition, near-to-far-field transformation, singular boundary condition, and simulation technique for the partially symmetric problem, followed by the parallel implementation of the BOR/FDTD method. Finally, two appendixes introduce the basic routines pertaining to the MPI library, its installation, configuration of this library and that of the MPICH, and how to set these up on a PC cluster.

We authors would like to thank our colleagues in the EMC Lab at Penn State for their help in testing the parallel code, especially Ji-Fu Ma, Neng-Tien Huang, Lai-Qing Ma, and Kai Du. We appriate Ph.D. student Johnathan N. Bringuier of the EMC Lab at Penn State for his help in proofreading the manuscript.

Finally, we would like to express our sincere gratitude to Cindy Zhou and Len Yu for their invaluable help during preparation of the manuscript.

Chapter 1

FDTD Method

As one of the major computational electromagnetics tools, the finite-difference time-domain (FDTD) method — originally proposed by K. S. Yee [1] — finds widespread use as a solver for a variety of electromagnetic problems. Although the focus of this book is on the parallelization of the FDTD method via the implementation of the message passing interface (MPI), we begin by reviewing some of the basic concepts of the FDTD in this chapter, including the finite-difference format, numerical dispersion, stability properties, and the FDTD update equations. Readers desiring to learn additional details about the FDTD method may refer to the literature [2–4] on this subject.

1.1 FINITE-DIFFERENCE CONCEPT

The FDTD method is a numerical technique based on the finite difference concept that is employed to solve Maxwell's equations for the electric and magnetic field distributions in both the time and spatial domains. Before introducing the FDTD method, we briefly review the concept of finite differencing. The increment of a piecewise function $f(x)$ at the point x can be expressed as:

$$\Delta f(x) = f(x + \Delta x) - f(x) \tag{1.1}$$

The difference quotient of the function $f(x)$ respect to the variable x can be written:

$$\frac{\Delta f(x)}{\Delta x} = \frac{f(x + \Delta x) - f(x)}{\Delta x} \tag{1.2}$$

Letting the increment Δx of the independent variable x approach 0, the derivative of the function $f(x)$ with respect to x may be written:

$$f'(x) = \lim_{\Delta x \to 0} \frac{\Delta f(x)}{\Delta x} = \lim_{\Delta x \to 0} \frac{f(x+\Delta x) - f(x)}{\Delta x} \qquad (1.3)$$

Thus, when the increment Δx is sufficiently small, the derivative of the function $f(x)$ with respect to x can be approximated by its difference quotient:

$$\frac{df}{dx} \approx \frac{\Delta f}{\Delta x} \qquad (1.4)$$

There are three alternate ways, defined below, by which we can express the derivative of the function $f(x)$: forward, backward, and the central difference. Their expressions read:

$$\left.\frac{\Delta f}{\Delta x}\right|_x = \frac{f(x+\Delta x) - f(x)}{\Delta x} \quad \text{(forward difference)} \qquad (1.5)$$

$$\left.\frac{\Delta f}{\Delta x}\right|_x = \frac{f(x) - f(x-\Delta x)}{\Delta x} \quad \text{(backward difference)} \qquad (1.6)$$

$$\left.\frac{\Delta f}{\Delta x}\right|_x = \frac{f(x+\Delta x/2) - f(x-\Delta x/2)}{\Delta x} \quad \text{(central difference)} \qquad (1.7)$$

Expressing the functions $f(x+\Delta x)$ and $f(x-\Delta x)$ at the point x in a Taylor series, we have:

$$f(x+\Delta x) = f(x) + \Delta x \frac{df(x)}{dx} + \frac{1}{2!}\Delta x^2 \frac{d^2 f(x)}{d^2 x} + \frac{1}{3!}\Delta x^3 \frac{d^3 f(x)}{d^3 x} \qquad (1.8)$$

$$f(x-\Delta x) = f(x) - \Delta x \frac{df(x)}{dx} + \frac{1}{2!}\Delta x^2 \frac{d^2 f(x)}{d^2 x} - \frac{1}{3!}\Delta x^3 \frac{d^3 f(x)}{d^3 x} \qquad (1.9)$$

Substituting (1.8) and (1.9) into (1.7), the central difference can be expressed in the following form:

$$\left.\frac{\Delta f}{\Delta x}\right|_x = \frac{df(x)}{dx} + \frac{1}{2^2}\frac{1}{3!}\Delta x^3 \frac{d^3 f(x)}{d^3 x} \qquad (1.10)$$

It is evident from (1.10) that the central difference is second-order accurate.

Similarly, substituting (1.8) and (1.9) into (1.5) and (1.6), we find that the forward and backward differences are only first-order accurate. Also the forward difference is always unstable, because it attempts to predict the future behavior of the function $f(x)$ from its current and previous values. The central difference, which is commonly used in the FDTD method, is conditionally stable. However, generally speaking, the backward difference is unconditionally stable, but leads to an implicit update procedure in which a matrix equation needs to be solved at each time step.

1.2 INTRODUCTION TO FDTD

The FDTD method utilizes the central difference approximation to discretize the two Maxwell's curl equations, namely, Faraday's and Ampere's laws, in both the time and spatial domains, and then solves the resulting equations numerically to derive the electric and magnetic field distributions at each time step using an explicit leapfrog scheme. The FDTD solution, thus derived, is second-order accurate, and is stable if the time step satisfies the Courant condition. One of the most important attributes of the FDTD algorithm is that it is embarrassingly parallel in nature because it only requires exchange of information at the interfaces of the subdomains in the parallel processing scheme. This will be evident when we discuss the details of the parallel FDTD technique in Chapter 7.

In Yee's scheme [1], the computational domain is discretized by using a rectangular grid. The electric fields are located along the edges of the electric elements, while the magnetic fields are sampled at the centers of the electric element surfaces and are oriented normal to these surfaces, this being consistent with the duality property of the electric and magnetic fields in Maxwell's equations. A typical electric element is shown in Figure 1.1(a), and the relative location of the electric and magnetic elements is also displayed in Figure 1.1(b). We will discuss only the electric elements throughout this book, though the properties of the magnetic elements are similar.

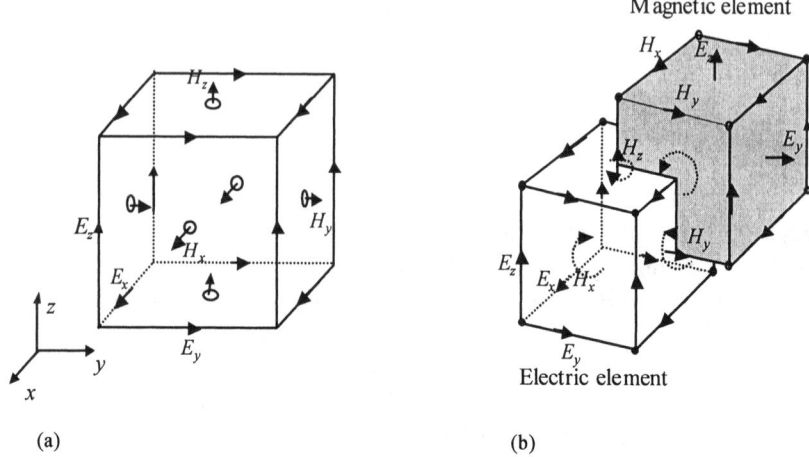

Figure 1.1 Position of the electric and magnetic fields in Yee's scheme. (a) Electric element. (b) Relationship between the electric and magnetic elements.

The FDTD utilizes rectangular pulses as basis functions in both the time and spatial domains, implying that the electric field is uniformly distributed along the edge of electric element, while the distribution of the magnetic fields is uniform on the surface of the electric element. In addition, in the time domain, the electric fields are sampled at times $n\Delta t$, and are assumed to be uniform in the time period of $(n-1/2\Delta t)$ to $(n+1/2\Delta t)$. Similarly, the magnetic fields are sampled at $(n+1/2\Delta t)$, and are assumed to be uniform in the time period of $n\Delta t$ to $(n+1)\Delta t$. The FDTD algorithm constructs a solution to the following two Maxwell's curl equations:

$$\nabla \times \vec{E} = -\mu \frac{\partial \vec{H}}{\partial t} - \sigma_M \vec{H} \qquad \text{(Faraday's law)} \qquad (1.11a)$$

$$\nabla \times \vec{H} = \varepsilon \frac{\partial \vec{E}}{\partial t} + \sigma \vec{E} \qquad \text{(Ampere's law)} \qquad (1.11b)$$

In the Cartesian coordinate system, we can rewrite (1.11a) and (1.11b) as the following six coupled partial differential equations:

$$\frac{\partial H_x}{\partial t} = \frac{1}{\mu_x}\left(\frac{\partial E_y}{\partial z} - \frac{\partial E_z}{\partial y} - \sigma_{Mx} H_x\right) \qquad (1.12a)$$

$$\frac{\partial H_y}{\partial t} = \frac{1}{\mu_y}\left(\frac{\partial E_z}{\partial x} - \frac{\partial E_x}{\partial z} - \sigma_{My} H_y\right) \quad (1.12b)$$

$$\frac{\partial H_z}{\partial t} = \frac{1}{\mu_z}\left(\frac{\partial E_x}{\partial y} - \frac{\partial E_y}{\partial x} - \sigma_{Mz} H_z\right) \quad (1.12c)$$

$$\frac{\partial E_x}{\partial t} = \frac{1}{\varepsilon_x}\left(\frac{\partial H_z}{\partial y} - \frac{\partial H_y}{\partial z} - \sigma_x E_x\right) \quad (1.12d)$$

$$\frac{\partial E_y}{\partial t} = \frac{1}{\varepsilon_y}\left(\frac{\partial H_x}{\partial z} - \frac{\partial H_z}{\partial x} - \sigma_y E_y\right) \quad (1.12e)$$

$$\frac{\partial E_z}{\partial t} = \frac{1}{\varepsilon_z}\left(\frac{\partial H_y}{\partial x} - \frac{\partial H_x}{\partial y} - \sigma_z E_z\right) \quad (1.12f)$$

where ε and σ, μ, and σ_M are the electric and magnetic parameters of the material, respectively. The anisotropic material can be described by using different values of dielectric parameters along the different direction. Equations (1.12a) to (1.12f) form the foundation of the FDTD algorithm for the modeling of interaction of the electromagnetic waves with arbitrary three-dimensional objects embedded in arbitrary media and excited by a given source (details of the source excitation can be found in Chapter 4). Using the conventional notations, the discretized fields in the time and spatial domains can be written in the following format:

$$E_x^n(i+1/2, j, k) = E_x\left((i+1/2)\Delta x, j\Delta y, k\Delta z, n\Delta t\right) \quad (1.13a)$$

$$E_y^n(i, j+1/2, k) = E_y\left(i\Delta x, (j+1/2)\Delta y, k\Delta z, n\Delta t\right) \quad (1.13b)$$

$$E_z^n(i, j, k+1/2) = E_z\left(i\Delta x, j\Delta y, (k+1/2)\Delta z, n\Delta t\right) \quad (1.13c)$$

$$\begin{aligned}H_x^{n+1/2}(i, j+1/2, k+1/2) \\ = H_x\left(i\Delta x, (j+1/2)\Delta y, (k+1/2)\Delta z, (n+1/2)\Delta t\right)\end{aligned} \quad (1.13d)$$

$$\begin{aligned}H_y^{n+1/2}(i+1/2, j, k+1/2) \\ = H_y\left((i+1/2)\Delta x, j\Delta y, (k+1/2)\Delta z, (n+1/2)\Delta t\right)\end{aligned} \quad (1.13e)$$

$$H_z^{n+1/2}(i+1/2, j+1/2, k) = H_z\left((i+1/2)\Delta x, (j+1/2)\Delta y, k\Delta z, (n+1/2)\Delta t\right) \tag{1.13f}$$

It is useful to note that the electric and magnetic fields in the discretized version are staggered in both time and space. For instance, the electric and magnetic fields are sampled at the time steps $n\Delta t$ and $(n+1/2)\Delta t$, respectively, and are also displaced from each other in space, as shown in Figure 1.1. Therefore, we need to interpolate the sampled electric and magnetic fields in order to measure the electric and magnetic fields in the continuous spatial and time domains. Ignoring this field sampling offset in the Fourier transforms may result in a significant error at high frequencies.

Using the notations in (1.13a) to (1.13f), we can represent Maxwell's equations (1.12a) to (1.12f) in the following explicit formats [2]:

$$H_x^{n+1/2}(i, j+1/2, k+1/2) = \frac{\mu_x - 0.5\Delta t \sigma_{Mx}}{\mu_x + 0.5\Delta t \sigma_{Mx}} H_x^{n-1/2}(i, j+1/2, k+1/2)$$

$$+ \frac{\Delta t}{\mu_x + 0.5\Delta t \sigma_{Mx}} \left[\begin{array}{c} \dfrac{E_y^n(i, j+1/2, k+1) - E_y^n(i, j+1/2, k)}{\Delta z} \\ - \dfrac{E_z^n(i, j+1, k+1/2) - E_z^n(i, j, k+1/2)}{\Delta y} \end{array} \right] \tag{1.14a}$$

$$H_y^{n+1/2}(i+1/2, j, k+1/2) = \frac{\mu_y - 0.5\Delta t \sigma_{My}}{\mu_y + 0.5\Delta t \sigma_{My}} H_y^{n-1/2}(i+1/2, j, k+1/2)$$

$$+ \frac{\Delta t}{\mu_y + 0.5\Delta t \sigma_{My}} \left[\begin{array}{c} \dfrac{E_z^n(i+1, j, k+1/2) - E_z^n(i, j, k+1/2)}{\Delta x} \\ - \dfrac{E_x^n(i+1/2, j, k+1) - E_x^n(i+1/2, j, k)}{\Delta z} \end{array} \right] \tag{1.14b}$$

$$H_z^{n+1/2}(i+1/2, j+1/2, k) = \frac{\mu_z - 0.5\Delta t \sigma_{Mz}}{\mu_z + 0.5\Delta t \sigma_{Mz}} H_z^{n-1/2}(i+1/2, j+1/2, k)$$

$$+ \frac{\Delta t}{\mu_z + 0.5\Delta t \sigma_{Mz}} \left[\begin{array}{c} \dfrac{E_x^n(i+1/2, j+1, k) - E_x^n(i+1/2, j, k)}{\Delta y} \\ - \dfrac{E_y^n(i+1, j+1/2, k) - E_y^n(i, j+1/2, k)}{\Delta x} \end{array} \right] \tag{1.14c}$$

$$E_x^{n+1}(i+1/2,j,k) = \frac{\varepsilon_x - 0.5\Delta t \sigma_x}{\varepsilon_x + 0.5\Delta t \sigma_x} E_x^n(i+1/2,j,k)$$

$$+ \frac{\Delta t}{\varepsilon_x + 0.5\Delta t \sigma_x} \left[\begin{array}{c} \dfrac{H_z^{n+1/2}(i+1/2,j+1/2,k) - H_z^{n+1/2}(i+1/2,j-1/2,k)}{\Delta y} \\ - \dfrac{H_y^{n+1/2}(i+1/2,j,k+1/2) - H_y^{n+1/2}(i+1/2,j,k-1/2)}{\Delta z} \end{array} \right] \quad (1.14d)$$

$$E_y^{n+1}(i,j+1/2,k) = \frac{\varepsilon_y - 0.5\Delta t \sigma_y}{\varepsilon_y + 0.5\Delta t \sigma_y} E_y^n(i,j+1/2,k)$$

$$+ \frac{\Delta t}{\varepsilon_y + 0.5\Delta t \sigma_y} \left[\begin{array}{c} \dfrac{H_x^{n+1/2}(i,j+1/2,k+1/2) - H_x^{n+1/2}(i,j+1/2,k-1/2)}{\Delta z} \\ - \dfrac{H_z^{n+1/2}(i+1/2,j+1/2,k) - H_z^{n+1/2}(i-1/2,j+1/2,k)}{\Delta x} \end{array} \right] \quad (1.14e)$$

$$E_z^{n+1}(i,j,k+1/2) = \frac{\varepsilon_z - 0.5\Delta t \sigma_z}{\varepsilon_z + 0.5\Delta t \sigma_z} E_z^n(i,j,k+1/2)$$

$$+ \frac{\Delta t}{\varepsilon_z + 0.5\Delta t \sigma_z} \left[\begin{array}{c} \dfrac{H_y^{n+1/2}(i+1/2,j,k+1/2) - H_y^{n+1/2}(i-1/2,j,k+1/2)}{\Delta x} \\ - \dfrac{H_x^{n+1/2}(i,j+1/2,k+1/2) - H_x^{n+1/2}(i,j-1/2,k+1/2)}{\Delta y} \end{array} \right] \quad (1.14f)$$

We point out that, for simplicity, we have omitted the explicit indices for the material parameters, which share the same indices with the corresponding field components. Equations (1.14a) through (1.14f) do not contain any explicit boundary information, and we need to augment them with an appropriate boundary condition in order to truncate the computational domain. In the FDTD simulation, some of the commonly used boundary conditions include those associated with the perfect electric conductor (PEC), the perfect magnetic conductor (PMC), the absorbing boundary condition (ABC), and the periodic boundary condition (PBC). In addition to the above boundary conditions, we also need to handle the interfaces between different media in an inhomogeneous environment. In accordance with the assumption of the locations of the electric and magnetic fields, the magnetic field is located along the line segment joining the two centers of adjacent cells. Consequently, the effective magnetic parameter corresponding to this magnetic field is the weighted average of the parameters of the material that fills the two adjacent cells. Unlike the magnetic field, the loop used to compute the electric field is likely to be distributed among four adjacent cells; therefore, the effective electric parameter corresponding to this electric field is equal to the weighted average of electric parameters of the material that fills these four cells. Also, the curved PEC and dielectric surfaces require the use of conformal FDTD technique [5, 6] for accurate modeling.

In recent years, research on the FDTD method has focused on the following three topics: improving the conventional FDTD algorithm and employing the conformal version instead in order to reduce the error introduced by the staircasing approximation; using a subgridding scheme in the FDTD technique to increase the local resolution [7–9]; and employing the alternative direction implicit (ADI) FDTD algorithm [10, 11] to increase the time step. In addition, new FDTD algorithms such as the multiresolution time domain (MRTD) method [12] and the pseudo-spectrum time domain (PSTD) technique [13] have been proposed with a view to lowering the spatial sampling. Yet another strategy, which has been found to be more robust as compared to the MRTD and PSTD, is to parallelize the conformal code [14, 15] and enhance it with either subgridding, the ADI algorithm, or both. The parallel FDTD algorithm gains the computational efficiency by distributing the burden on a cluster. It also enables one to solve large problems that could be beyond the scope of a single processor because of CPU time limitations.

1.3 NUMERICAL DISPERSION

If a medium is dispersive, then the propagation velocities of electromagnetic waves will vary with frequency in such a medium. In a nondispersive medium, for instance, in free space, the radian frequency and the wave numbers satisfy the relationship:

$$\left(\frac{\omega}{c}\right)^2 = k_x^2 + k_y^2 + k_z^2 \qquad (1.15)$$

where k_x, k_y, and k_z are the propagation constants along the x-, y-, and z-directions, respectively. Even when the medium is nondispersive, electromagnetic waves inside the FDTD mesh travel along different directions at different speeds, and this phenomenon is referred to as the so-called numerical dispersion error. This error is a function of the FDTD cell size, its shape as well as that of the differencing format used to discretize the differential equations.

We will now proceed to investigate the numerical dispersion characteristics of the FDTD method in the Cartesian coordinate system. We begin by representing a plane wave function as:

$$\psi(x,y,z,t) = \psi_0 \exp\left[j(\omega t - k_x x - k_y y - k_z z)\right] \tag{1.16}$$

where the radian frequency $\omega = 2\pi f$. If the discretizations in the x-, y-, and z-directions and the time step are Δx, Δy, Δz, and Δt, respectively, the expression in (1.16) can be written as:

$$\psi^n(I,J,K) = \psi_0 \exp\left[j(\omega n\Delta t - k_x I\Delta x - k_y J\Delta y - k_z K\Delta z)\right] \tag{1.17}$$

where n, I, J, and K are the indices in time and space, respectively. In free space, electromagnetic waves satisfy the following wave equation:

$$\left(\frac{\partial^2}{\partial x^2} + \frac{\partial^2}{\partial y^2} + \frac{\partial^2}{\partial z^2} - \frac{1}{c^2}\frac{\partial^2}{\partial t^2}\right)\psi = 0 \tag{1.18}$$

where c is the velocity of light in free space. Using the central differencing in both the time and spatial domains, we can discretize (1.18) to get:

$$\begin{aligned}
&\frac{\psi^n(i+1,j,k) - 2\psi^n(i,j,k) + \psi^n(i-1,j,k)}{\Delta x^2} \\
&+ \frac{\psi^n(i,j+1,k) - 2\psi^n(i,j,k) + \psi^n(i,j-1,k)}{\Delta y^2} \\
&+ \frac{\psi^n(i,j,k+1) - 2\psi^n(i,j,k) + \psi^n(i,j,k-1)}{\Delta z^2} \\
&= \frac{\psi^{n+1}(i,j,k) - 2\psi^n(i,j,k) + \psi^{n-1}(i,j,k)}{\Delta t^2}
\end{aligned} \tag{1.19}$$

Substituting (1.17) into (1.19), we have:

$$\begin{aligned}
&\left[\frac{1}{c\Delta t}\sin\left(\frac{\omega\Delta t}{2}\right)\right]^2 \\
&= \left[\frac{1}{\Delta x}\sin\left(\frac{k_x\Delta x}{2}\right)\right]^2 + \left[\frac{1}{\Delta y}\sin\left(\frac{k_y\Delta y}{2}\right)\right]^2 + \left[\frac{1}{\Delta z}\sin\left(\frac{k_z\Delta z}{2}\right)\right]^2
\end{aligned} \tag{1.20}$$

It is evident that (1.20) reduces to (1.15) as $\Delta x \to 0$, $\Delta y \to 0$, and $\Delta z \to 0$, implying that the numerical dispersion error decreases when the cell size becomes smaller. In addition, the numerical dispersion error is different along different propagation directions, and we will now illustrate this phenomenon through a simple two-dimensional example of propagation in a uniform mesh (i.e., for $\Delta x = \Delta y$). We assume that a line source, oriented along the z-direction, is located in free space. In Figure 1.2, we plot the distribution of its radiated field E_z in the x-y plane. The white circle in the figure depicts an ideal wave front when there is no dispersion. It is evident that the numerical dispersion error is smaller along the diagonals in $\phi = 45°$, $135°$, $225°$, and $315°$ than in other directions. In addition, we note that the numerical dispersion error becomes worse as the cell size is increased progressively.

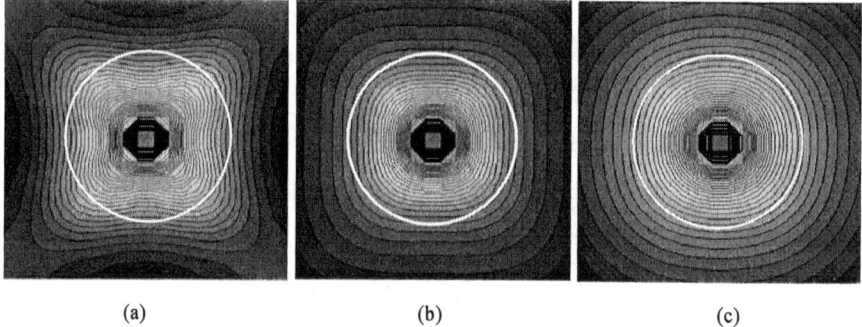

Figure 1.2 Numerical dispersive error in the FDTD method. (a) $\Delta x = \Delta y = \lambda/5$. (b) $\Delta x = \Delta y = \lambda/10$. (c) $\Delta x = \Delta y = \lambda/20$.

1.4 STABILITY ANALYSIS

One of the critical issues that we must address when developing a code that utilizes the marching-on-time technique is the stability of the algorithm. The stability characteristic of the FDTD algorithm depends upon the nature of the physical model, differencing technique employed, and the quality of the mesh structure. To understand the nature of the stability characteristic, we solve (1.20) for ω to get:

$$\omega = \frac{2}{\Delta t} \sin^{-1}\left(c\Delta t \sqrt{\frac{1}{\Delta x^2}\sin^2\left(\frac{k_x \Delta x}{2}\right) + \frac{1}{\Delta y^2}\sin^2\left(\frac{k_y \Delta y}{2}\right) + \frac{1}{\Delta z^2}\sin^2\left(\frac{k_z \Delta z}{2}\right)} \right) \quad (1.21)$$

If ω is an imaginary number, we know from (1.16) that the electromagnetic waves either will attenuate rapidly to zero, or will grow exponentially and become divergent, depending on whether the imaginary part is positive or negative. In order to ensure that ω is a real number instead, the expression inside the round bracket in (1.21) must satisfy the condition:

$$c\Delta t \sqrt{\frac{1}{\Delta x^2}\sin^2\left(\frac{k_x\Delta x}{2}\right)+\frac{1}{\Delta y^2}\sin^2\left(\frac{k_y\Delta y}{2}\right)+\frac{1}{\Delta z^2}\sin^2\left(\frac{k_z\Delta z}{2}\right)} \leq 1 \quad (1.22)$$

Since the maximum possible value of the sine-square term under the square root is 1, the time step must satisfy:

$$\Delta t \leq \frac{1}{c\sqrt{\frac{1}{\Delta x^2}+\frac{1}{\Delta y^2}+\frac{1}{\Delta z^2}}} \quad (1.23)$$

in order for the solution to be stable. The criterion above is called the stability condition for the FDTD method, and it is referred to as Courant condition (or the Courant, Friedrichs, and Lewy criterion) [16]. Equation (1.23) indicates that the time step is determined by the cell sizes in the x-, y-, and z-directions, and the speed of light in the medium.

To help the reader gain further physical insight into (1.23), we simplify (1.23) to a 1-D case, where the Courant condition is simplified to $c\Delta t \leq \Delta x$. The time required for the field to propagate from the nth to $(n+1)$th node is obviously $\Delta t = \Delta x/c$. In the FDTD simulations, let us suppose we choose a time step $\Delta t > \Delta x/c$. Then, the algorithm yields a nonzero field value before the wave can reach the $(n+1)$th node, traveling at the speed of light. This would violate causality and result in an unstable solution.

1.5 NONUNIFORM MESH

As mentioned earlier, the central difference that is typically used for the FDTD simulation is second-order accurate for a uniform mesh. Suppose, however, we use a nonuniform mesh in which the cell sizes deviate from a constant, at least in one direction, as shown in Figure 1.3. Then the electric field update equation will lose its second-order accuracy, because this field would no longer be located at the center of two adjacent magnetic field points.

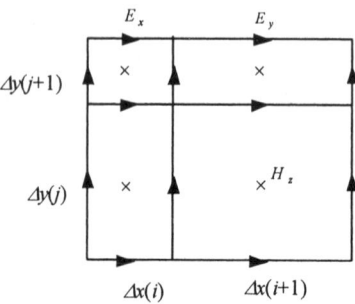

Figure 1.3 Typical nonuniform mesh in the FDTD simulation.

Alternatively, we can minimize this loss in accuracy in nonuniform meshing, and yet retain the advantage of reducing the memory requirement and simulation time if we modify the update equation for this case in a manner explained below. To reduce the error caused by the nonuniform mesh, we should avoid any abrupt discontinuities in the cell size, and let them vary relatively smoothly instead. Typically, we generate the grid to conform to the specified geometry and employ a nonuniform grid such that the line locations coincide with the interface. For example, when designing an FDTD mesh for the simulation of a patch antenna, we first line up the mesh grids with the surfaces of the dielectric substrate, and then design the mesh distribution inside the slab. A finer mesh is used in the region where the fields are expected to vary rapidly.

Next, we present the field-update scheme for the nonuniform FDTD method that is designed to mitigate the loss of the second-order accuracy. We note, first of all, that since the magnetic fields are located at the centers of the FDTD grids, and their calculation from (1.14a) to (1.14c) satisfies the central differencing scheme, hence, they are second-order accurate. We now modify the update equations for the electric fields as follows. For the E_x component, for instance, we modify (1.14d) as:

$$E_x^{n+1}(i+1/2,j,k) = \frac{\varepsilon_x - 0.5\Delta t \sigma_x}{\varepsilon_x + 0.5\Delta t \sigma_x} E_x^n(i+1/2,j,k) + \frac{\Delta t}{\varepsilon_x + 0.5\Delta t \sigma_x}$$
$$\times \left[\frac{H_z^{n+1/2}(i+1/2,j+1/2,k) - H_z^{n+1/2}(i+1/2,j-1/2,k)}{0.5(\Delta y(j) + \Delta y(j-1))} - \frac{H_y^{n+1/2}(i+1/2,j,k+1/2) - H_y^{n+1/2}(i+1/2,j,k-1/2)}{0.5(\Delta z(k) + \Delta z(k-1))} \right] \quad (1.24)$$

The accuracy of the field E_x depends on the choice of the ratio between the two adjacent cell sizes. If this ratio is smaller than 1.2, then the result derived by using (1.24) would be quite accurate, despite the loss of the second-order accuracy.

The modification on the electric field calculation affects the accuracy of the subsequent magnetic field calculations. Therefore, we need slightly to modify the magnetic field update equation, for instance, the H_z-component can be written as:

$$H_z^{n+1/2}(i+1/2, j+1/2, k) = \frac{\mu_z - 0.5\Delta t \sigma_{Mz}}{\mu_z + 0.5\Delta t \sigma_{Mz}} \times H_z^{n-1/2}(i+1/2, j+1/2, k)$$

$$+ \frac{\Delta t}{\mu_z + 0.5\Delta t \sigma_{Mz}} \begin{bmatrix} \dfrac{E_x^n(i+1/2, j+1, k) - E_x^n(i+1/2, j, k)}{0.25(\Delta y(j-1) + 2\Delta y(j) + \Delta y(j+1))} - \\ \dfrac{E_y^n(i+1, j+1/2, k) - E_y^n(i, j+1/2, k)}{0.25(\Delta x(i-1) + 2\Delta x(i) + \Delta x(i+1))} \end{bmatrix} \quad (1.25)$$

We can go one step further [17] and modify the denominator of (1.25) so that the update equation reads as follows:

$$H_z^{n+1/2}(i+1/2, j+1/2, k) = \frac{\mu_z - 0.5\Delta t \sigma_{Mz}}{\mu_z + 0.5\Delta t \sigma_{Mz}} \times H_z^{n-1/2}(i+1/2, j+1/2, k)$$

$$+ \frac{\Delta t}{\mu_z + 0.5\Delta t \sigma_{Mz}} \begin{bmatrix} \dfrac{E_x^n(i+1/2, j+1, k) - E_x^n(i+1/2, j, k)}{0.125(\Delta y(j-1) + 6\Delta y(j) + \Delta y(j+1))} - \\ \dfrac{E_y^n(i+1, j+1/2, k) - E_y^n(i, j+1/2, k)}{0.125(\Delta x(i-1) + 6\Delta x(i) + \Delta x(i+1))} \end{bmatrix} \quad (1.26)$$

Experience shows that the use of (1.26) typically improves the accuracy over that of (1.25). Besides putting a cap on the ratio of two adjacent cell sizes, we also recommend that the ratio of the maximum to the minimum cell size in one direction be less than 20, to control the error resulting from the effects of the deviation from a uniform mesh size.

REFERENCES

[1] K. Yee, "Numerical Solution of Initial Boundary Value Problems Involving Maxwell's Equations in Isotropic Media," *IEEE Transactions on Antennas and Propagation*, Vol. 14, No. 5, May 1966, pp. 302-307.

[2] A. Taflove and S. Hagness, *Computational Electromagnetics: The Finite-Difference Time-Domain Method*, 3rd ed., Artech House, Norwood, MA, 2005.

[3] W. Yu and R. Mittra, *Conformal Finite-Difference Time-Domain Maxwell's Equations Solver: Software and User's Guide*, Artech House, Norwood, MA, 2004.

[4] W. Yu, et al., *Parallel Finite Difference Time Domain*, Press of Communication University of China, Beijing, 2005, (in Chinese).

[5] W. Yu and R. Mittra, "A Conformal FDTD Software Package for Modeling of Antennas and Microstrip Circuit Components," *IEEE Antennas and Propagation Magazine*, Vol. 42, No. 5, October 2000, pp. 28-39.

[6] W. Yu and R. Mittra, "A Conformal Finite Difference Time Domain Technique for Modeling Curved Dielectric Surfaces," *IEEE Microwave and Guided Wave Letters*, Vol. 11, No. 1, January 2001, pp. 25-27.

[7] B. Wang, et al., "A Hybrid 2-D ADI-FDTD Subgridding Scheme for Modeling On-Chip Interconnects," *IEEE Transactions on Advanced Packaging*, Vol. 24, No. 11, November 2001, pp. 528-533.

[8] W. Yu and R. Mittra, "A New Subgridding Method for Finite Difference Time Domain (FDTD) Algorithm," *Microwave and Optical Technology Letters*, Vol. 21, No. 5, June 1999, pp. 330-333.

[9] M. Marrone, R. Mittra, and W. Yu, "A Novel Approach to Deriving a Stable Hybrid FDTD Algorithm Using the Cell Method," *Proc. IEEE AP-S URSI*, Columbus, OH, 2003.

[10] T. Namiki, "A New FDTD Algorithm Based on Alternating-Direction Implicit Method," *IEEE Transactions on Microwave Theory and Techniques*, Vol. 47, No. 10, October 1999, pp. 2003-2007.

[11] F. Zheng, Z. Chen, and J. Zhang, "Toward the Development of a Three-Dimensional Unconditionally Stable Finite-Difference Time-Domain Method," *IEEE Transactions on Microwave Theory and Techniques*, Vol. 48, No. 9, September 2000, pp. 1550-1558.

[12] Y. Chao, Q. Cao, and R. Mittra, *Multiresolution Time Domain Scheme for Electromagnetic Engineering*, John Wiley & Sons, New York, 2005.

[13] Q. Liu, "The PSTD Algorithm: A Time-Domain Method Requiring Only Two Cells Per Wavelength," *Microwave and Optical Technology Letters*, Vol. 15, 1997, pp. 158-165.

[14] W. Yu, et al., "A Robust Parallelized Conformal Finite Difference Time Domain Field Solver Package Using the MPI Library," *IEEE Antennas and Propagation Magazine*, Vol. 47, No. 3, March 2005, pp. 39-59.

[15] C. Guiffaut and K. Mahdjoubi, "A Parallel FDTD Algorithm Using the MPI Library," *IEEE Antennas and Propagation Magazine*, Vol. 43, No. 2, April 2001, pp. 94-103.

[16] R. Courant, K. Friedrichs, and H. Lewy, "Uber die partiellen Differenzengleichungen der math_ematischen Physik," *Math, An.*, Vol. 100, 1928, pp. 32-74.

[17] W. Yu and R. Mittra, "A Technique of Improving the Accuracy of the Nonuniform Time-Domain Algorithm," *IEEE Transactions on Microwave Theory and Techniques*, Vol. 47, No. 3, March 1999, pp. 353-356.

Chapter 2

Boundary Conditions

It is well known that boundary conditions play a very important role in FDTD simulations, because they are used to truncate the computational domain when modeling an open region problem. Though the original FDTD algorithm was proposed as early as 1966, it was not really used to solve practical problems until the early 1980s when Mur's absorbing boundary [1] was proposed. Though Mur's absorbing boundary condition is relatively simple, and while it has been successfully used to solve many engineering problems, it has room for improvement in terms of the accuracy of the solution it generates. To improve its accuracy, Mei and Fang [2] have introduced the so-called super absorption technique, while Chew [3] has proposed to employ Liao's boundary condition — both of which exhibit better characteristics than that of Mur, especially for obliquely incident waves. However, many of these absorbing boundary conditions were found to suffer from either an instability problem or inaccurate solution, and the quest for robust and effective boundary conditions continued unabated until the perfectly matched layer (PML) was introduced by Berenger [4]. In contrast to the other boundary conditions such as those of Mur and Liao, an infinite PML can absorb the incoming waves at all frequencies as well as for all incident angles. In this chapter, we first discuss several different boundary conditions for mesh truncation in the FDTD simulations, including the perfect electric conductor (PEC), the perfect magnetic conductor (PMC), and the first-order Mur's absorbing boundary condition (ABC). Following this, we proceed to discuss the PML boundary condition in detail. Although we refer to the PML as a type of ABC, it is in fact an anisotropic material — albeit mathematical — which is inserted in the periphery of the computational domain in order to absorb the outgoing waves.

Before introducing the boundary conditions, we first investigate their role in the FDTD simulations. In the FDTD, Maxwell's equations that govern the relationship between the electric and magnetic fields in the time and spatial domains are translated into difference equations, which do not explicitly contain any boundary information. It is necessary, therefore, to combine the difference equations with the appropriate boundary conditions in order to carry out the mesh truncation as a preamble to solving these equations. Generally speaking, there are two types of boundary conditions that we require in the FDTD simulations: the interface condition between different media, and the outer boundary condition for mesh truncation. In this chapter, we only discuss the latter, namely, the boundary that is used to truncate the computational domain.

Let us consider a typical FDTD mesh, as shown in Figure 2.1, which also displays the field locations. The electric and magnetic fields inside the computational domain can be computed through the FDTD update equations that work with the field values at previous time steps at these locations, as well as those at the nearest neighbors. This is true except for the electric fields on the boundaries, because the magnetic field values outside the domain needed for the electric field updating are not available there. The task of the boundary condition is to obtain the required electric field values by using those available in the interior region.

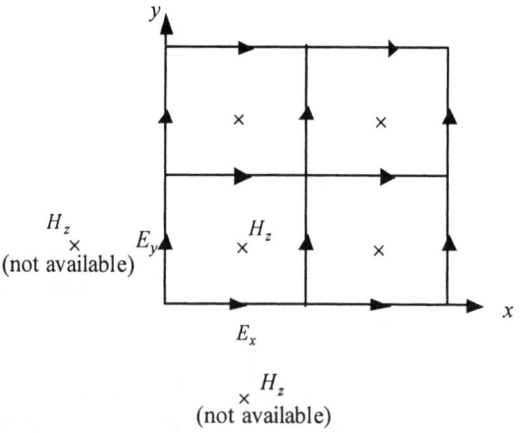

Figure 2.1 Field distribution in the x-y plane.

2.1 PEC AND PMC BOUNDARY CONDITIONS

The perfect electric conductor (PEC) is a natural boundary for electromagnetic waves, since it totally reflects the waves falling upon it. When the PEC condition is applied to truncate the FDTD computational domain, it simply forces the tangential electric fields on the domain boundary to be zero. Examples of the PEC are waveguide and cavity walls, and the ground plane of a microwave circuit or a patch antenna with ideally infinite conductivity. For an odd symmetric system, the PEC can be employed to truncate the FDTD domain at the plane of symmetry to reduce the size of the computational domain. Once we know the field value inside the computational domain, we can obtain the corresponding values in the region of symmetry via the use of the image theory. In addition, the PEC can be also used to truncate periodic structures for a normally incident plane wave.

In common with the PEC, the perfect magnetic conductor (PMC) also is a natural type of boundary condition for electromagnetic waves, and it is also totally reflecting. However, unlike the PEC, the PMC is not physical, but it is merely an artifice. Both the PEC and PMC are often used to take advantage of the symmetry of the object geometry with a view to reducing the size of the computational domain, or to truncate a periodic structure for the normally incident plane wave case. In principle, we can use the PMC boundary to truncate a computational domain by imposing the condition that forces the tangential magnetic fields on the PMC surface to be zero. However, since the tangential magnetic fields are located at the centers of the FDTD cells, their positions vary with the FDTD cell size. This uncertainty associated with the PMC boundary location makes it slightly difficult to impose the PMC boundary condition. In the FDTD simulations, we typically place the PMC boundaries such that they coincide with the domain boundary, but do not impose it on outermost tangential magnetic fields that have a half-cell offset from the domain boundary. We implement the PMC boundary through the image theory, satisfied by the magnetic fields at both sides of the PMC boundary, as shown in Figure 2.2.

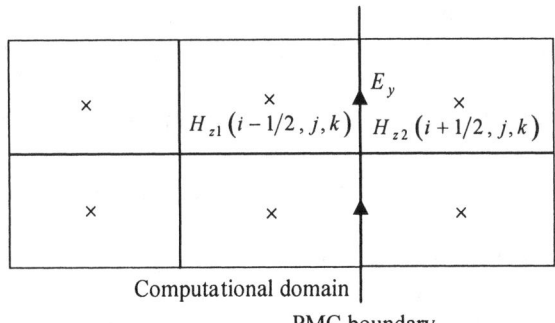

Figure 2.2 PMC boundary condition.

For the PMC setting as shown in Figure 2.2, the tangential electric fields are still located at the PMC boundary. They are calculated by using the tangential magnetic fields at both sides of the PMC boundary and the normal component of the magnetic field on the PMC boundary. The virtual magnetic field outside the PMC boundary can be obtained from the image theory and its use enables us to follow the usual scheme to update the electric fields on the PMC boundary. Referring to Figure 2.2, we know from the image theory that the magnetic field H_{z2} outside the PMC boundary is equal to negative value of H_{z1}, and is located at the image point of H_{z1}. Thus, within the computational domain we can set $H_z^{n+1/2}(i+1/2, j+1/2, k) = -H_z^{n+1/2}(i-1/2, j+1/2, k)$, and proceed to update the tangential electric fields on the PMC boundary as follows:

$$E_y^{n+1}(i, j+1/2, k) = \frac{\varepsilon_y - 0.5\Delta t \sigma_y}{\varepsilon_y + 0.5\Delta t \sigma_y} E_y^n(i, j+1/2, k)$$

$$+ \frac{\Delta t}{\varepsilon_y + 0.5\Delta t \sigma_y} \left(\frac{\Delta H_x^{n+1/2}}{0.5(\Delta z(k-1) + \Delta z(k))} + \frac{2H_z^{n+1/2}(i-1/2, j+1/2, k)}{0.5(\Delta x(i-1) + \Delta x(i))} \right) \quad (2.1)$$

From (2.1), we note that the computation of the tangential electric fields E_y on the PMC boundary not only requires the magnetic fields outside the computational domain, but also the normal component of the magnetic fields at the PMC boundary; consequently, we also need to update the normal component of the magnetic field, which is not required in conventional FDTD simulations. To do this, we need to allocate one extra cell for the magnetic field arrays when programming the update equations, and let the magnetic fields in the outermost layer of the computational domain serve the role of the PMC. Neither the PEC nor PMC boundary condition requires any information exchange between the processes, and hence, they do not need special treatment in the parallel processing.

2.2 MUR'S ABSORBING BOUNDARY CONDITION

Prior to the advent of the PML, Mur's ABC played an important role in the development of FDTD. Even today, for objects such as waveguides, patch antennas, and microwave circuits, we can still utilize this boundary condition to achieve a reasonably good result in the FDTD simulations, but without paying the heavy price exacted by the other types of ABCs. Mur's ABC has advantages over the PML in terms of simulation speed and memory requirement. In this book, we only discuss the first-order Mur's boundary condition since we choose the PML for most applications that require high accuracy. Although higher-order Mur's ABCs are superior to their first-order counterpart, their implementation is more complex and is not well-suited for parallel processing because they require exchange of field information that places a heavier burden in terms of communication than do either the first-order Mur or the PML, which are the preferred choices for parallel implementation.

Returning to Mur's ABC, we point out that it can be viewed as an approximation of the Engquist-Majda boundary condition [5], which we will briefly review next for the one-dimensional case. Let us suppose that there is a plane wave that propagates along the x-direction. The field variable $\phi(x+ct)$, associated with this wave, satisfies the following equation:

$$\left(\frac{\partial}{\partial x} - \frac{1}{c}\frac{\partial}{\partial t}\right)\phi(x,t) = 0 \tag{2.2}$$

whose solution can be written as $\phi(x+ct)$, when the wave propagates along the negative x-direction, as shown in Figure 2.3. Referring to Figure 2.3, we observe that the field inside the domain, say at a point P, can be updated through the regular FDTD equations as formulated in (2.2). However, the field at a point on the boundary, say at Q, has to be updated by using an ABC.

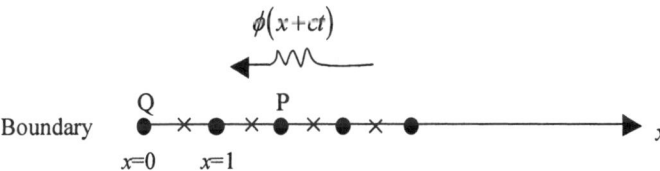

Figure 2.3 Absorbing boundary condition.

Since no information as to the behavior of the field is available outside the domain $(x<0)$, the Engquist-Majda absorbing boundary condition is based on expressing the field at the boundary of the domain in terms of the known fields in the interior of the domain. To develop this ABC, we return to (2.2) — which is exact for the normally-incident plane wave case — and use the forward difference formula to approximate it in both the time and spatial domains. We can then express the field at the domain boundary $x = 0$, and at time $(n+1)\Delta t$, as:

$$\phi_{x=0}^{n+1} = \left(1 - \frac{c\Delta t}{\Delta x}\right)\phi_{x=0}^{n} + \frac{c\Delta t}{\Delta x}\phi_{x=1}^{n} \tag{2.3}$$

From (2.3) we observe that the field at the domain boundary can be represented as a summation of two terms, namely, the fields at the boundary and adjacent to the boundary, both sampled at the previous time step. Since (2.2) is a 1-D wave equation, the expression in (2.3) is only valid for a normally incident plane wave. Mur's ABC is based on approximating (2.2) by using central differencing in both the time and spatial domains to get:

$$\frac{1}{\Delta x}\left[\phi_{x=1}^{n+1/2} - \phi_{x=0}^{n+1/2}\right] = \frac{1}{c\Delta t}\left[\phi_{x=1/2}^{n+1} - \phi_{x=1/2}^{n}\right] \quad (2.4)$$

It is obvious that (2.4) is second-order accurate; however, it neither includes the field at the boundary of the domain, nor at time $(n + 1)\Delta t$. Mur proceeds to handle this problem by representing both $\phi_{x=0}^{n+1/2}$ and $\phi_{x=1/2}^{n}$ in terms of averages of the two adjacent fields in both the time and spatial domains, as follows:

$$\phi_{x=0}^{n+1/2} = \frac{1}{2}\left(\phi_{x=0}^{n+1} + \phi_{x=0}^{n}\right) \quad (2.5a)$$

$$\phi_{x=1/2}^{n} = \frac{1}{2}\left(\phi_{x=1}^{n} + \phi_{x=0}^{n}\right) \quad (2.5b)$$

Next, substitution of the (2.5a) and (2.5b) into (2.4), followed by some simplifications leads us to the so-called Mur's first-order boundary condition, which reads:

$$\phi_{x=0}^{n+1} = \phi_{x=1}^{n} + \frac{c\Delta t - \Delta x}{c\Delta t + \Delta x}\left[\phi_{x=1}^{n+1} - \phi_{x=0}^{n}\right] \quad (2.6)$$

In practical FDTD simulations, we can usually obtain acceptable results by using the first-order Mur's absorbing boundary condition, given in (2.6), which is relatively simple if the simulated objects and excitation sources are not too close to the domain boundary. On the left domain boundary, the tangential electric fields E_y and E_z can be expressed as:

$$E_y^{n+1}(0, j+1/2, k) = E_y^{n}(1, j+1/2, k) + \frac{c\Delta t - \Delta x(0)}{c\Delta t + \Delta x(0)}\left[E_y^{n+1}(1, j+1/2, k) - E_y^{n}(0, j+1/2, k)\right] \quad (2.7)$$

$$E_z^{n+1}(0, j, k+1/2) = E_z^n(1, j, k+1/2)$$
$$+ \frac{c\Delta t - \Delta x(0)}{c\Delta t + \Delta x(0)} \left[E_z^{n+1}(1, j, k+1/2) - E_z^n(0, j, k+1/2) \right] \quad (2.8)$$

where c is the velocity of light in free space, and $\Delta x(0)$ is the cell size adjacent to the domain boundary. If the material parameter of the medium is different from that of free space, the velocity of light c appearing in (2.7) and (2.8) should be replaced by $c/\sqrt{\varepsilon_r}$, where ε_r is the relative dielectric constant of the medium. We note from (2.7) and (2.8) that the tangential field components E_y and E_z are associated with the cell size $\Delta x(0)$, and the time step Δt. Because of this, we have found that when using the Mur's ABC, the cell with the minimum size should always be located at the boundary of the domain to avoid spurious reflections. Because the computation of the fields E_y and E_z at the domain boundary only requires the knowledge of previous values at the same locations and the field values adjacent to the domain boundary at the same time step, all the information needed for parallel processing is available in the individual subdomains; hence, the first-order Mur's ABC is easily incorporated in a code designed for parallel processing. We should point out that (2.2) is only first-order accurate for 2-D and 3-D problems, and that its error increases with the angle of the incidence wave impinging upon the boundary. Hence, its use is not recommended when we are dealing with propagation of waves at highly oblique angles, or when a highly accurate solution is desired.

2.3 UNSPLIT PML

As mentioned earlier, the PML is really an artificial anisotropic material that can be perfectly matched to free space at all incident angles and frequencies, provided that the interface is an infinite plane. Though a variety of different versions have been proposed in the literature since the original one was introduced by Berenger [4], the central concept embodied in these alternative PMLs is still the same as found in [6]; however, the different versions do lead to different codes when implemented in the FDTD.

There are four variables appearing in Maxwell's equations, namely, fields, material parameters, coordinates, and time. We point out that even though the FDTD can simulate static problems and can derive the static charge distribution, this quantity does not explicitly appear in the FDTD update equations. The original PML, created by Berenger, describes the matched anisotropic material through a split-field representation of Maxwell's equations. Rather than using split-field PML, Gedney [7] works directly with the anisotropic material and incorporates them into Maxwell's equations. Chew [8, 9] and Rapaport [10] follow a slightly different strategy and utilizes the stretched coordinate system, as well as complex coordinates, to describe the anisotropic material. In common with Gedney's unsplit PML (UPML), Chew's PML does not require the field splitting inside the PML either. Among the different versions of the PML, the UPML formulation is the easiest one to understand, and has been broadly applied in the FDTD simulations. One of the principal advantages of using the UPML in an FDTD code is that Maxwell's equations have the same form in both the PML region and the computational domain, and hence, we can employ a coding strategy that remains consistent as we go from one region to the other.

Assuming that the time dependence is given by $e^{j\omega t}$, Maxwell's equations in a general anisotropic medium can be expressed as:

$$\nabla \times \tilde{\vec{E}} = -j\omega\mu\bar{\bar{s}}\tilde{\vec{H}} \tag{2.9a}$$

$$\nabla \times \tilde{\vec{H}} = j\omega\varepsilon\bar{\bar{s}}\tilde{\vec{E}} \tag{2.9b}$$

where $\bar{\bar{s}}$ is a tensor given by:

$$\bar{\bar{s}} = \begin{bmatrix} S_y S_z S_x^{-1} & 0 & 0 \\ 0 & S_x S_z S_y^{-1} & 0 \\ 0 & 0 & S_x S_y S_z^{-1} \end{bmatrix} \tag{2.10}$$

In the above equations, $S_x = 1 + \dfrac{\sigma_{x,\text{PML}}}{j\omega\varepsilon_0}$, $S_y = 1 + \dfrac{\sigma_{y,\text{PML}}}{j\omega\varepsilon_0}$, $S_z = 1 + \dfrac{\sigma_{z,\text{PML}}}{j\omega\varepsilon_0}$. Also, $\sigma_{x,\text{PML}}$, $\sigma_{y,\text{PML}}$, and $\sigma_{z,\text{PML}}$ are the conductivity distributions inside the PML region that are equal to zero in the computational domain not contained within the PML layer. Thus, we have $S_x = S_y = S_z = 1$ in the non-PML region, and (2.9a) and (2.9b) revert to conventional Maxwell's equations in this region. Substituting (2.10) into (2.9a) and (2.9b), we obtain, for the E_x component:

$$\varepsilon_x \left(1 + \dfrac{\sigma_x}{j\omega\varepsilon_x}\right) \dfrac{S_y S_z}{S_x} \tilde{E}_x = \dfrac{\partial \tilde{H}_z}{\partial y} - \dfrac{\partial \tilde{H}_y}{\partial z} \tag{2.11}$$

where ε_x and σ_x are the electrical parameters of the materials in the computational domain. Our objective is to transform (2.11) into the time domain and then represent it as an explicit update equation. The presence of the frequency factors at the left-hand side of (2.11) makes it difficult for us to express it directly in the time domain via Fourier transforms. To circumvent this difficulty, we express (2.11) as an explicit time-update equation by introducing two intermediate variables. We first define the variable $\tilde{P}_x$, which is related to $\tilde{E}_x$ as follows:

$$\varepsilon_x \left(1 + \frac{\sigma_x}{j\omega \varepsilon_x}\right) \tilde{E}_x = \tilde{P}_x \tag{2.12}$$

Taking the inverse Fourier transform of (2.12), we obtain the following equation:

$$\varepsilon_x \frac{\partial E_x}{\partial t} + \sigma_x E_x = P_x \tag{2.13}$$

in the time domain. Next, substituting (2.12) into (2.11), we have:

$$\frac{S_y S_z}{S_x} \tilde{P}_x = \frac{\partial \tilde{H}_z}{\partial y} - \frac{\partial \tilde{H}_y}{\partial z} \tag{2.14}$$

We now define another intermediate variable $\tilde{Q}_x$ satisfying the relationship:

$$\frac{S_y}{S_x} \tilde{P}_x = \tilde{Q}_x \tag{2.15}$$

Substituting S_x and S_y into (2.15), we have:

$$\left(1 + \frac{\sigma_{y,\text{PML}}}{j\omega \varepsilon_0}\right) \tilde{P}_x = \left(1 + \frac{\sigma_{x,\text{PML}}}{j\omega \varepsilon_0}\right) \tilde{Q}_x \tag{2.16}$$

Next, simplifying (2.16) we obtain:

$$\left(j\omega \varepsilon_0 + \sigma_{y,\text{PML}}\right) \tilde{P}_x = \left(j\omega \varepsilon_0 + \sigma_{x,\text{PML}}\right) \tilde{Q}_x \tag{2.17}$$

Taking the inverse Fourier transform of (2.17), we get:

$$\left(\varepsilon_0 \frac{\partial P_x}{\partial t} + \sigma_{y,\text{PML}} P_x\right) = \left(\varepsilon_0 \frac{\partial Q_x}{\partial t} + \sigma_{x,\text{PML}} Q_x\right) \quad (2.18)$$

Substituting the intermediate variable $\tilde{Q}_x$ into (2.14), we have:

$$S_z \tilde{Q}_x = \frac{\partial \tilde{H}_z}{\partial y} - \frac{\partial \tilde{H}_y}{\partial z} \quad (2.19)$$

Substituting the expression of S_z into (2.19), we derive:

$$\left(1 + \frac{\sigma_{z,\text{PML}}}{j\omega\varepsilon_0}\right)\tilde{Q}_x = \frac{\partial \tilde{H}_z}{\partial y} - \frac{\partial \tilde{H}_y}{\partial z} \quad (2.20)$$

Rewriting (2.20) as:

$$\left(j\omega + \frac{\sigma_{z,\text{PML}}}{\varepsilon_0}\right)\tilde{Q}_x = j\omega \frac{\partial \tilde{H}_z}{\partial y} - j\omega \frac{\partial \tilde{H}_y}{\partial z} \quad (2.21)$$

and then taking the inverse Fourier transform of (2.21), we finally get:

$$\left(\frac{\partial Q_x}{\partial t} + \frac{\sigma_{z,\text{PML}}}{\varepsilon_0} Q_x\right) = \frac{\partial H_z}{\partial y} - \frac{\partial H_y}{\partial z} \quad (2.22)$$

Equation (2.22) has a format that is similar to that of the conventional FDTD equation, and Q_x can be solved using this equation together with a leapfrog updating scheme. Once we have obtained Q_x, we can solve for the intermediate variable P_x from it using (2.18), and then derive for E_x from P_x through (2.13). We note that using the above procedure for the computation of the electric field inside the PML region requires three time updates for lossy materials, together with the use of two intermediate variables. The magnetic field computation inside the PML region requires a similar procedure.

Note that the conductivity distributions $\sigma_{x,\text{PML}}$ and $\sigma_{z,\text{PML}}$ only exist inside the PML region, and $\sigma_{x,\text{PML}}$, $\sigma_{y,\text{PML}}$, and $\sigma_{z,\text{PML}}$ are 1-D functions of x, y, and z, respectively. In addition, we note that the anisotropic material in the PML region has characteristics that are different from those of conventional dielectric materials that are isotropic. For instance, $\sigma_{x,\text{PML}}$ is not only located at the position of E_x, but it also affects the other field components. In addition, $\sigma_{x,\text{PML}}$ is nonzero only when it is perpendicular to the PML layer, and is zero when it is parallel. The same statement is true for the edge and corner regions because they can be viewed as the combination of two or three PML regions. It is suggested that the reader refer to [7] for further details pertaining to the characteristics of PML layers.

The absorption characteristics of the PML, described above, are not sufficiently good for surface waves and at low frequencies. In order to improve the performance for the two cases mentioned above, we will introduce the time convolution PML [11], in which the parameters S_s ($s = x, y, z$) will be modified to a slightly different format.

An ideal PML layer described by [7] is reflectionless if the material in the PML region satisfies $\sigma_{\text{PML}}/\varepsilon_0 = \sigma_{M,\text{PML}}/\mu_0$. If both the σ_{PML} and $\sigma_{M,\text{PML}}$ distributions in the PML region are properly selected, the power into the PML region will be dissipated due to the conductivity loss. However, a single-layer PML cannot absorb the power penetrating into the PML region perfectly, because of the numerical approximations in the FDTD that simulate the performance of the PML layer. To overcome this problem, it becomes necessary to use a multilayer PML instead. Suppose x is the distance measured from the outer boundary of the computational domain, then the conductivity distribution in the PML region is given by:

$$\sigma(x) = \sigma_{\max}\left(\frac{d-x}{d}\right)^m \tag{2.23}$$

where the index m is taken to be either 2 or 4. Also, in (2.23), d is the thickness of the PML region and $\sigma_{\max}$ is the maximum value of the conductivity, which can be expressed as:

$$\sigma_{\max} = \frac{m+1}{200\pi\sqrt{\varepsilon_r}\Delta x} \tag{2.24}$$

Even though the above strategy leads to a PML that exhibits good performance, this type of unsplit PML has two disadvantages: first, the associated updated equation depends on the simulated material; second, its reflection characteristics at low frequencies are relatively poor. Below we will introduce two alternative PML types that do not depend on the characteristics of the simulated materials, and hence, are more robust.

Before closing this section, we point out that the two interim variables in the update procedure described above, P and Q, are located at the same points as the corresponding field components, and hence, we do not need to pass on any extra information for the PML implementation in parallel processing.

2.4 STRETCHED COORDINATE PML

We will discuss the stretched coordinate PML, which was proposed by Chew and Wood in 1994 [8] and independently by Rapaport [10]. The unsplit PML is implemented by modifying the material parameters in Maxwell's equations, the update equations are different for different materials; for example, the update procedure for the lossy materials is more complex than that for lossless materials. The stretched coordinate PML introduces a coordinate-stretching instead of simulating anisotropic materials inside the PML region; hence, it is independent of the materials being simulated.

Maxwell's equations in the stretched coordinate system can be written in the format:

$$\nabla_\sigma \times \tilde{\vec{E}} = -j\omega\mu\tilde{\vec{H}} \qquad (2.25a)$$

$$\nabla_\sigma \times \tilde{\vec{H}} = j\omega\varepsilon\tilde{\vec{E}} \qquad (2.25b)$$

$$\nabla_\sigma \cdot \varepsilon\tilde{\vec{E}} = \rho \qquad (2.25c)$$

$$\nabla_\sigma \times \mu\tilde{\vec{H}} = 0 \qquad (2.25d)$$

where $\nabla_\sigma = \hat{x}\dfrac{1}{S_x}\dfrac{\partial}{\partial x} + \hat{y}\dfrac{1}{S_y}\dfrac{\partial}{\partial y} + \hat{z}\dfrac{1}{S_z}\dfrac{\partial}{\partial z}$, and S_x, S_y, and S_z are identical to those employed in Section 2.3. It can be shown that the stretched coordinate PML is mathematically equivalent to the unsplit PML.

The above approach has recently been further developed by Chew, Jin, and Michielssen to yield the so-called complex coordinate PML [9], in which the complex coordinates are obtained via the following variable transformations:

$$\bar{x} = \int_0^x S_x(x')\,dx' \qquad (2.26a)$$

$$\bar{y} = \int_0^y S_y(y')\,dy' \qquad (2.26b)$$

$$\bar{z} = \int_0^z S_z(z')\,dz' \qquad (2.26c)$$

Using these, the differential operator ∇_σ in (2.25) can be written as $\bar{\nabla} = \hat{x}\frac{\partial}{\partial \bar{x}} + \hat{y}\frac{\partial}{\partial \bar{y}} + \hat{z}\frac{\partial}{\partial \bar{z}}$. By choosing these complex variables $\bar{x}$, $\bar{y}$, and $\bar{z}$, the wave in the PML region can be made to decay. Furthermore, we have the desired property that there are no reflections from the interface between the PML region and the computational domain. An important advantage of the stretched and complex coordinate PMLs is that they are relatively easy to implement in different coordinate systems. Suppose x is distance measured from the outer boundary of the computational domain, the S_x in the PML region will be expressed as:

$$S_x(x) = 1 - j\alpha\left(\frac{d-x}{d}\right)^2 \qquad (2.27)$$

and $S_x(x) = 1$ inside the computational domain, where α is a real number greater than zero. The low reflection characteristics of this PML can be realized by selecting an appropriate α.

Using (2.26a), the complex coordinate $\bar{x}$ in the PML region can be written as:

$$\bar{x} = 1 + j\alpha \frac{(d-x)^3}{3d^2} \qquad (2.28)$$

and $\bar{x} = 1$ inside the computational domain. As pointed out in the last section, we must use a multilayer PML to reduce the numerical errors, and a six-layer PML is a good practical choice for most problems.

2.5 TIME CONVOLUTION PML

The time convolution PML [11] is based on the stretched coordinate PML [8, 10], which was described in the last section. The six coupled Maxwell's equations in the stretched coordinate PML can be written in the following form:

$$j\omega\varepsilon\tilde{E}_x + \sigma_x\tilde{E}_x = \frac{1}{S_y}\frac{\partial \tilde{H}_z}{\partial y} - \frac{1}{S_z}\frac{\partial \tilde{H}_y}{\partial z} \qquad (2.29a)$$

$$j\omega\varepsilon\tilde{E}_y + \sigma_y\tilde{E}_y = \frac{1}{S_z}\frac{\partial \tilde{H}_x}{\partial z} - \frac{1}{S_x}\frac{\partial \tilde{H}_z}{\partial x} \qquad (2.29b)$$

$$j\omega\varepsilon\tilde{E}_z + \sigma_z\tilde{E}_z = \frac{1}{S_x}\frac{\partial \tilde{H}_y}{\partial x} - \frac{1}{S_y}\frac{\partial \tilde{H}_x}{\partial y} \qquad (2.29c)$$

$$j\omega\mu_x\tilde{H}_x + \sigma_{Mx}\tilde{H}_x = \frac{1}{S_z}\frac{\partial \tilde{E}_y}{\partial z} - \frac{1}{S_y}\frac{\partial \tilde{E}_z}{\partial y} \qquad (2.29d)$$

$$j\omega\mu_y\tilde{H}_y + \sigma_{My}\tilde{H}_y = \frac{1}{S_x}\frac{\partial \tilde{E}_z}{\partial x} - \frac{1}{S_z}\frac{\partial \tilde{E}_x}{\partial z} \qquad (2.29e)$$

$$j\omega\mu_z\tilde{H}_z + \sigma_{Mz}\tilde{H}_z = \frac{1}{S_y}\frac{\partial \tilde{E}_x}{\partial y} - \frac{1}{S_x}\frac{\partial \tilde{E}_y}{\partial x} \qquad (2.29f)$$

To derive the update equations for the time convolution PML from (2.29a), we first take its Laplace transform to obtain the following equation in the time domain:

$$\varepsilon_x \frac{\partial E_x}{\partial t} + \sigma_x E_x = \overline{S}_y(t) * \frac{\partial H_z}{\partial y} - \overline{S}_z(t) * \frac{\partial H_y}{\partial z} \qquad (2.30)$$

where $\overline{S}_y$ and $\overline{S}_z$ are the Laplace transforms of $1/S_y$ and $1/S_z$, respectively.

The time convolution PML is derived by converting (2.30) to a form that is suitable for explicit updating. Furthermore, to overcome the shortcomings of the split field and unsplit PMLs insofar as the effectiveness at the low frequencies and the absorption of the surface waves are concerned, we modify S_x, S_y, and S_z as follows [12]:

$$S_x = K_x + \frac{\sigma_{x,\text{PML}}}{\alpha_x + j\omega\varepsilon_0}, \quad S_y = K_y + \frac{\sigma_{y,\text{PML}}}{\alpha_y + j\omega\varepsilon_0}, \quad S_z = K_z + \frac{\sigma_{z,\text{PML}}}{\alpha_z + j\omega\varepsilon_0}$$

where $\alpha_{x,y,z}$ and $\sigma_{x,y,z,\text{PML}}$ are real numbers, and K is greater than 1. $\overline{S}_x$, $\overline{S}_y$, and $\overline{S}_z$ can be obtained from the Laplace transforms:

$$\overline{S}_x = \frac{\delta(t)}{K_x} - \frac{\sigma_x}{\varepsilon_0 K_x} \exp\left[-\left(\sigma_{x,\text{PML}}/\varepsilon_0 K_x + \alpha_{x,\text{PML}}/\varepsilon_0\right) tu(t)\right] = \frac{\delta(t)}{K_x} + \xi_x(t) \quad (2.31)$$

$$\overline{S}_y = \frac{\delta(t)}{K_y} - \frac{\sigma_y}{\varepsilon_0 K_y} \exp\left[-\left(\sigma_{y,\text{PML}}/\varepsilon_0 K_y + \alpha_{y,\text{PML}}/\varepsilon_0\right) tu(t)\right] = \frac{\delta(t)}{K_y} + \xi_y(t) \quad (2.32)$$

$$\overline{S}_z = \frac{\delta(t)}{K_z} - \frac{\sigma_z}{\varepsilon_0 K_z} \exp\left[-\left(\sigma_{z,\text{PML}}/\varepsilon_0 K_z + \alpha_{z,\text{PML}}/\varepsilon_0\right) tu(t)\right] = \frac{\delta(t)}{K_z} + \xi_z(t) \quad (2.33)$$

where $\delta(t)$ and $u(t)$ are an impulse function and a step function, respectively. Substituting (2.32) and (2.33) into (2.30), we have:

$$\varepsilon_x \varepsilon_0 \frac{\partial E_x}{\partial t} + \sigma_x E_x = \frac{1}{K_y}\frac{\partial H_z}{\partial y} - \frac{1}{K_z}\frac{\partial H_y}{\partial z} + \xi_y(t) * \frac{\partial H_z}{\partial y} - \xi_z(t) * \frac{\partial H_y}{\partial z} \quad (2.34)$$

It is not numerically efficient to compute the convolution directly appearing in (2.34), and to address this issue we introduce a quantity $Z_{0y}(m)$, in order to calculate it efficiently, as follows:

$$\begin{aligned}Z_{0y}(m) &= \int_{m\Delta t}^{(m+1)\Delta t} \xi_y(\tau)d\tau \\ &= -\frac{\sigma_y}{\varepsilon_0 K_y^2}\int_{m\Delta t}^{(m+1)\Delta t}\exp\left[-\left(\sigma_{y,\mathrm{PML}}\Big/\varepsilon_0 K_y + \alpha_{y,\mathrm{PML}}\Big/\varepsilon_0\right)\tau\right]d\tau \\ &= a_y \exp\left[-\left(\sigma_{y,\mathrm{PML}}\Big/K_y + \alpha_{y,\mathrm{PML}}\right)\left(m\Delta t\Big/\varepsilon_0\right)\right]\end{aligned} \quad (2.35)$$

where

$$a_y = \frac{\sigma_{y,\mathrm{PML}}}{\sigma_{y,\mathrm{PML}}K_y + K_y^2\alpha_y}\left(\exp\left[-\left(\sigma_{y,\mathrm{PML}}\Big/K_y + \alpha_y\right)\left(m\Delta t\Big/\varepsilon_0\right)\right]-1\right) \quad (2.36)$$

A similar expression can be derived for $Z_{0z}(m)$. Using (2.35) and (2.36), (2.34) can be written as:

$$\begin{aligned}\varepsilon_x\varepsilon_0 &\frac{E_x^{n+1}(i+1/2,j,k)-E_x^n(i+1/2,j,k)}{\Delta t} + \sigma_x \frac{E_x^{n+1}(i+1/2,j,k)+E_x^n(i+1/2,j,k)}{2} \\ &= \frac{H_z^{n+1/2}(i+1/2,j+1/2,k)-H_z^{n+1/2}(i+1/2,j-1/2,k)}{K_y\Delta y} \\ &\quad - \frac{H_y^{n+1/2}(i+1/2,j,k+1/2)-H_y^{n+1/2}(i+1/2,j,k-1/2)}{K_z\Delta z} \\ &\quad + \sum_{m=0}^{N-1}Z_{0y}(m)\frac{H_z^{n-m+1/2}(i+1/2,j+1/2,k)-H_z^{n-m+1/2}(i+1/2,j-1/2,k)}{K_y\Delta y} \\ &\quad - \sum_{m=0}^{N-1}Z_{0z}(m)\frac{H_y^{n-m+1/2}(i+1/2,j,k+1/2)-H_y^{n-m+1/2}(i+1/2,j,k-1/2)}{K_z\Delta z}\end{aligned} \quad (2.37)$$

Finally, the update formula of (2.37) takes the form:

$$\varepsilon_x \varepsilon_0 \frac{E_x^{n+1}(i+1/2,j,k) - E_x^n(i+1/2,j,k)}{\Delta t} + \sigma_x \frac{E_x^{n+1}(i+1/2,j,k) + E_x^n(i+1/2,j,k)}{2}$$

$$= \frac{H_z^{n+1/2}(i+1/2,j+1/2,k) - H_z^{n+1/2}(i+1/2,j-1/2,k)}{K_y \Delta y} \quad (2.38)$$

$$- \frac{H_y^{n+1/2}(i+1/2,j,k+1/2) - H_y^{n+1/2}(i+1/2,j,k-1/2)}{K_z \Delta z}$$

$$+ \psi_{exy}^{n+1/2}(i+1/2,j,k) - \psi_{exz}^{n+1/2}(i+1/2,j,k)$$

where,

$$\psi_{exy}^{n+1/2}(i+1/2,j,k) = b_y \psi_{exy}^{n-1/2}(i+1/2,j,k)$$
$$+ a_y \frac{H_z^{n+1/2}(i+1/2,j+1/2,k) - H_z^{n+1/2}(i+1/2,j-1/2,k)}{\Delta y} \quad (2.39)$$

$$\psi_{exz}^{n+1/2}(i+1/2,j,k) = b_z \psi_{exz}^{n-1/2}(i+1/2,j,k)$$
$$+ a_z \frac{H_y^{n+1/2}(i+1/2,j,k+1/2) - H_y^{n+1/2}(i+1/2,j,k-1/2)}{\Delta z} \quad (2.40)$$

$$b_x = \exp\left[-\left(\sigma_{x,\text{PML}}/K_x + \alpha_x\right)\left(\Delta t/\varepsilon_0\right)\right] \quad (2.41a)$$

$$b_y = \exp\left[-\left(\sigma_{y,\text{PML}}/K_y + \alpha_y\right)\left(\Delta t/\varepsilon_0\right)\right] \quad (2.41b)$$

$$b_z = \exp\left[-\left(\sigma_{z,\text{PML}}/K_z + \alpha_z\right)\left(\Delta t/\varepsilon_0\right)\right] \quad (2.41c)$$

Equation (2.38) is the desirable update equation that we have been seeking to retain the advantages of the unsplit PML, and overcome its drawback at the same time. In common with the conventional FDTD, the electric field updating inside the PML region only requires the magnetic fields around it and the value of ψ at the previous time step. The same statement is true for the magnetic field update as well. The time domain convolution PML does not require additional information exchange in the parallel FDTD simulation, over and above that in the conventional FDTD. The conductivity distribution in the PML region can be expressed as (2.23). Suppose that y is a distance measured from the outer boundary of the computational domain, then the distribution of K_y is given by:

$$K_y(y) = 1 + (K_{max} - 1)\frac{|d-y|^m}{d^m} \tag{2.42}$$

The implementation of the time convolution PML in the FDTD code is relatively simpler than most of the other types of PMLs. Also, this type of PML does not depend on the properties of the materials being simulated. In addition, it has a good performance at low frequencies.

Numerical experiments have proven that the performance of the PML is not very sensitive to the choice of K. For example, we do not observe any significant variation in the reflection property of the PML when K_{max} is varied from 1 to 15. However, the value of α in (2.36) and (2.41) plays an important role in determining the reflection level of the time domain convolution PML. We have found, by carrying out numerical experiments, that the value of α to be chosen not only depends on the cell size inside the PML region but also on the width of excitation pulse used in the FDTD simulation; furthermore, it is not possible to employ a single α value to cover all the possible cell sizes and pulse shapes that we might use. Fortunately, however, we can easily find a suitable value of α for a narrow range of cell size and pulse shape. For example, for a Gaussian pulse modulated with a sinusoidal function (f_{3dB} = 1 GHz and $f_{modulation}$ = 1 GHz), the highest frequency of spectrum of the excitation pulse is about 5 GHz, and the cell sizes in the FDTD simulations are selected to be 0.0001 m to 0.006 m. For this case, the value of α can be chosen as follows:

$$\alpha_x = \eta \frac{0.06}{300\pi\Delta x}, \alpha_y = \eta \frac{0.06}{300\pi\Delta y}, \alpha_z = \eta \frac{0.06}{300\pi\Delta z} \tag{2.43}$$

where $\eta = 1$. We should point out that if we change the width of the excitation pulse while keeping the cell size unchanged, there will be significant reflection from the PML layer, and therefore the value of α should be a function of the width of the excitation pulse. A recommended choice for α, expressed in terms of η, is as follows:

$$\eta = \begin{cases} \dfrac{1}{1+0.6(\lambda_{min} - \lambda_{reference})/\lambda_{reference}}, & \lambda_{min} > \lambda_{reference} \\ 1 & \lambda_{min} = \lambda_{reference} \\ \dfrac{1}{1+(\lambda_{reference} - \lambda_{min})/\lambda_{reference}}, & \lambda_{min} < \lambda_{reference} \end{cases} \tag{2.44}$$

where λ_{min} is the shortest wavelength in the spectrum of the reference excitation pulse, and λ_{min} is the shortest wavelength in the FDTD simulations. It is useful to point out that we have introduced α in the time convolution PML to improve performance at the lower frequencies. The reflection level of this PML will revert to the same as that of others if we set $\alpha = 0$.

The most important features of the time convolution PML are that its update procedure is not related to the properties of the material that fills in the computational domain and that its performance is good even at low frequencies.

REFERENCES

[1] G. Mur, "Absorbing Boundary Conditions for the Finite-Difference Approximation of the Time-Domain Electromagnetic Field Equations," *IEEE Transactions on Electromagnetic Compatibility*, Vol. 23, No. 3, 1981, pp. 377-382.

[2] K. Mei and J. Fang, "Superabsorption —— A Method to Improve Absorbing Boundary Conditions," *IEEE Transactions on Antennas and Propagation*, Vol. 40, No. 9, September 1992, pp. 1001-1010.

[3] Z. Liao, et al., "A Transmitting Boundary for Transient Wave Analyses," *Scientia Sinica*, (Series A), Vol. 27, No. 10, October 1984, pp. 1062-1076.

[4] J. Berenger, "A Perfectly Matched Layer Medium for the Absorption of Electromagnetic Waves," *J. Comput.*, Vol. 114, October 1994, pp. 185-200.

[5] B. Engquist and A. Majda, "Absorbing Boundary Conditions for the Numerical Simulation of Waves," *Mathematics of Computation*, Vol. 31, 1977, pp. 629-651.

[6] D. Werner and R. Mittra, "A New Field Scaling Interpretation of Berenger's PML and Its Comparison to Other PML Formulations," *Microwave and Optical Technology Letters*, Vol. 16, No. 2, 1997, pp. 103-106.

[7] S. Gedney, "An Anisotropic Perfectly Matched Layer-Absorbing Medium for the Truncation of FDTD Lattices," *IEEE Transactions on Antennas and Propagation*, Vol. 44, No. 12, December 1996, pp. 1630-1639.

[8] W. Chew and W. Wood, "A 3-D Perfectly Matched Medium from Modified Maxwell's Equations with Stretched Coordinates," *Microwave and Optical Technology Letters*, Vol. 7, 1994, pp. 599-604.

[9] W. Chew, J. Jin, and E. Michielssen, "Complex Coordinate Stretching as a Generalized Absorbing Boundary Condition," *Microwave and Optical Technology Letters*, Vol. 15, No. 6, August 1997, pp. 363-369.

[10] C. Rapaport, "Perfectly Matched Absorbing Boundary Conditions Based on Anisotropic Lossy Mapping of Space," *IEEE Microwave and Guided Wave Letters*, Vol. 5, 1995, pp. 90-92.

[11] J. Roden and S. Gedney, "Convolution PML (CPML): An Efficient FDTD Implementation of the CFS-PML for Arbitrary Medium," *Microwave and Optical Technology Letters*, Vol. 27, No. 5, 2000, pp. 334-339.

[12] M. Kuzuoglu and R. Mittra, "Frequency Dependence of the Constitutive Parameters of Causal Perfectly Matched Anisotropic Absorbers," *IEEE Microwave and Guided Wave Letters*, Vol. 6, No. 6, 1996, pp. 447-449.

Chapter 3

Improvement of the FDTD

In principle, the conventional FDTD algorithm implemented on a Cartesian grid can be used to solve almost all electromagnetic modeling problems, provided we are willing and able to use sufficiently fine discretization to model the geometry and material characteristics of the object accurately. However, for many real-world problems it is often necessary to employ enhanced versions of the FDTD in order to render the problem manageable and treatable with the available resource within a reasonable time frame and without placing an unrealistic burden on the CPU. In this chapter, we discuss some of these techniques for enhancing the conventional FDTD method, including conformal techniques, implicit updating schemes, treatment of dispersive media, and handling of lumped elements. The use of these techniques enables us to handle many real-world problems that would otherwise not be amenable to analysis using the conventional FDTD.

3.1 CONFORMAL FDTD FOR PEC OBJECTS

One of the most important drawbacks of the conventional FDTD method stems from the fact that it uses a staircasing approximation to model objects whose geometries do not conform to the Cartesian grid employed in the FDTD algorithm. On the other hand, the Cartesian grid makes the algorithm easier to implement and thus more attractive. Theoretically, the error introduced by a staircasing approximation reduces as the cell size is reduced; however, such a move dramatically increases the burden on the computer memory and causes the simulation time to increase proportionately. An attempt to increase the fineness of the discretization to achieve the desired accuracy may be totally impractical for the available computational resource. In this section, we will discuss the modeling of curve PEC objects using a modification of the FDTD algorithm that is not based on increasing the mesh density to mitigate the staircasing problem. We begin by reviewing the history of conformal FDTD techniques for PEC objects [1–4] that continue to utilize the framework of the Cartesian grid with little compromise of the accuracy.

In a typical staircasing approximation, an FDTD cell is entirely filled with the object material if the center of the FDTD cell is embedded inside the object; otherwise, it is regarded as though it was completely outside the object, as shown in Figure 3.1. The level of actual staircasing error depends on the mesh distribution, shape of the cell, and geometry of the object. Generally speaking, staircasing error can be neglected when the cell size is less than $\lambda/60$. However, the conformal FDTD algorithm that will soon be presented will continue the use of $\lambda/20$-size cells, which is standing for modeling objects that conform precisely to a Cartesian FDTD grid.

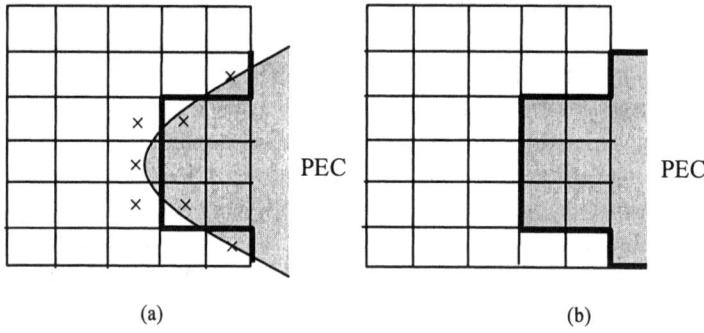

Figure 3.1 Staircasing approximation. (a) Original geometry. (b) Staircasing approximation.

The simplest conformal technique is based on a diagonal approximation, in which the cells partially filled with PEC are approximated by a diagonal line, as shown in Figure 3.2.

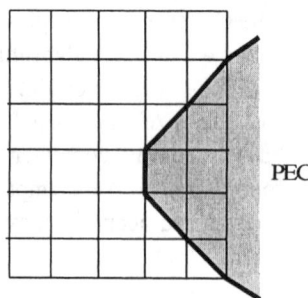

Figure 3.2 Diagonal approximation.

We can see from Figure 3.2 that some of the cells would now be half-filled with PEC. In the conventional FDTD update equations, two of the electric field components in such cells reside inside the PEC object, and are set to zero. Also, when computing the magnetic fields, only the two nonzero electric fields contribute to the update equation. At the same time, the areas of the Faraday loop in the magnetic field calculation are halved. One consequence of this is that the time step size in the FDTD simulation must also be halved to ensure the stability of FDTD solution. Though relative simple, the simple approximation yields reasonably accurate results in many cases — but not always.

In 1992, Jurgens et al. proposed the contour-path method [1], in which the Faraday loop for the magnetic field update is made to be a part of the cell outside the PEC object, as shown in Figure 3.3. This approach is more accurate than the diagonal approximation though more complex. In addition, the approach requires a reduction of the time step from that employed in the staircased version, since some of the partially filled cells are relatively smaller. To avoid dealing with extremely small partially filled cells that require a drastic reduction of the time step, these cells are combined with their neighboring ones. Although it enables one to increase the time step, it makes the algorithm even more involved in terms of its implementation because it requires a significant modification of the field update equations in the deformed cells, and frequently leads to an unstable solution.

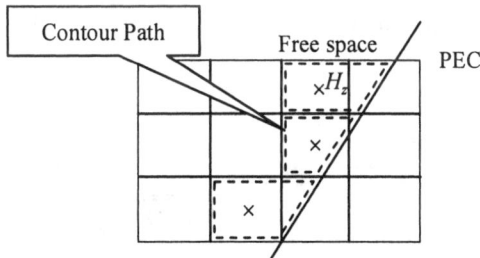

Figure 3.3 Contour path conformal technique.

In 1997, Dey and Mittra proposed a new technique [2] for handling objects that do not conform to Cartesian grids, which requires only a modification of the Faraday loop, as shown in Figure 3.4. In this method, a conformal updating algorithm is applied to individual partially filled cells, and the time step size needs to be reduced in accordance with the size and shape of the deformed cell. Although the integration path may be outside the object, the sampling point of the magnetic field is always assumed to be at the center of the FDTD cell. Although this method is usually quite accurate, the reduction of the time step could be considerable if the area of the Faraday loop is small. Also the approach may suffer from late time instabilities when modeling infinite thin PEC structures, and hence, it is not widely used in practice.

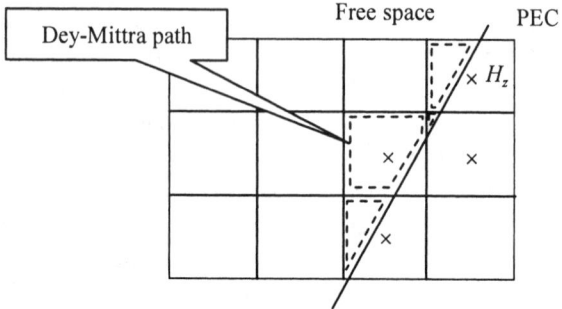

Figure 3.4 Modified Faraday loop in the Dey-Mittra technique.

The major difficulty in the above approaches is the need to reduce the time step when there are integration loops present anywhere in the computational domain, in order to avoid the occurrence of late time instabilities. We now proceed to describe a slight modification of the Dey-Mittra conformal scheme that mitigates the problem of late time instability and guarantees a stable solution. The Faraday loop for this algorithm introduced by Yu and Mittra [3, 4] are sketched in Figure 3.5.

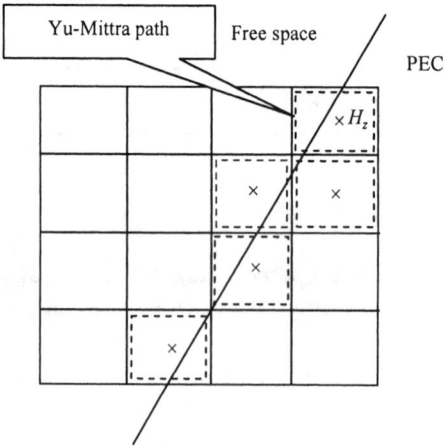

Figure 3.5 Faraday loop of another conformal technique.

To overcome the time step reduction problem, we use the same integration path (see Figure 3.6) as that in the conventional FDTD. The magnetic fields are still sampled at the centers of the FDTD cells, and the electric fields inside the PEC object do not contribute to the magnetic field. Since in this scheme we neither change the magnetic field sampling points nor the pattern of the Faraday loops, the electric field update equations remain the same as that in the conventional FDTD. Now the magnetic field update equation becomes:

$$H_z^{n+1/2}(i+1/2,j+1/2,k) = H_z^{n-1/2}(i+1/2,j+1/2,k) \\ + \frac{\Delta t}{\mu_z} \left[\begin{array}{c} \dfrac{\Delta x_0 E_x^n(i+1/2,j+1,k) - \Delta x E_x^n(i+1/2,j,k)}{\Delta x_0 \Delta y_0} \\ - \dfrac{\Delta y_0 E_y^n(i+1,j+1/2,k) - \Delta y E_y^n(i,j+1/2,k)}{\Delta x_0 \Delta y_0} \end{array} \right] \quad (3.1)$$

The variables appearing in the above equation are defined in Figure 3.6. Since the areas of the Faraday loops are the same as those in a conventional grid, unlike other schemes the time step in FDTD simulation does not have to be changed. Though the accuracy may be compensated slightly in this scheme as compared to the previous two methods, the robustness of the method, as well as the fact it does not require a reduction in the time step, makes it very attractive for solving practical problems.

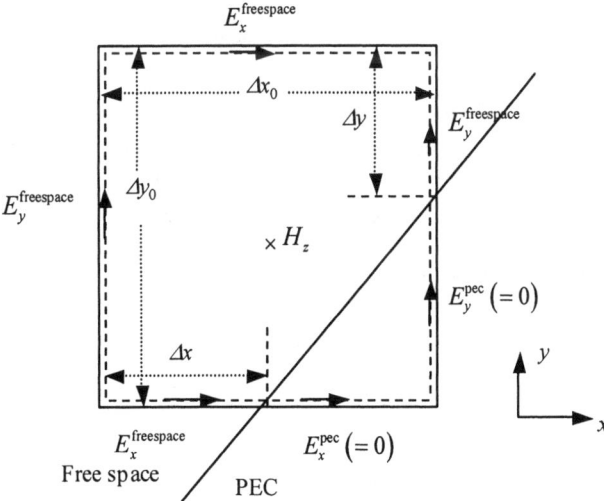

Figure 3.6 Integration path used in Yu-Mittra conformal technique.

Next, we explain how the Yu-Mittra conformal technique is applied to model the curved or slanted thin PEC objects. If the FDTD cell size is sufficiently small, the curved PEC surface inside that cell can be approximated by a planer facet, as shown in Figure 3.7. We see from this figure that the thin PEC splits one FDTD cell into two; consequently, since the two sides of this surface are isolated, there exist two separate magnetic fields H_z inside the deformed cell and two sets of electric fields on the edges truncated by the thin PEC surface. We will now describe the field update schemes when such a surface is present in the computational domain.

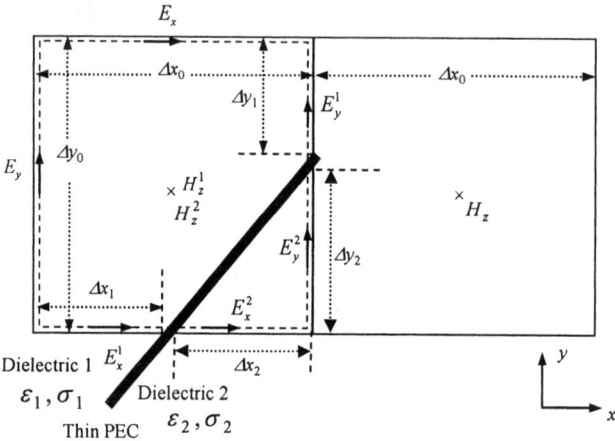

Figure 3.7 Integration path for thin PEC in Yu-Mittra conformal technique.

While H_z^1 and H_z^2 belong to the two nonoverlapping regions containing the both dielectrics 1 and 2, both of the H-fields are still considered as being collocated at the center of the cell for the purpose of FDTD updating. The electric fields corresponding to H_z^1 are E_y^1, E_x^1, E_x, and E_y, and the field H_z^1 can be updated by using the formula:

$$H_z^{1,n+1/2}(i+1/2, j+1/2, k) = H_z^{1,n-1/2}(i+1/2, j+1/2, k) \\ + \frac{\Delta t}{\mu_z}\left[\frac{\Delta x_0 E_x^n(i+1/2, j+1, k) - \Delta x_1 E_x^{1,n}(i+1/2, j, k)}{\Delta x_0 \Delta y_0} \\ - \frac{\Delta y_1 E_y^{1,n}(i+1, j+1/2, k) - \Delta y_0 E_y^n(i, j+1/2, k)}{\Delta x_0 \Delta y_0}\right] \quad (3.2)$$

Likewise, the update equation for H_z^2 involves the electric fields E_x^2 and E_y^2, and the magnetic field can be updated by using the formula:

$$H_z^{2,n+1/2}(i+1/2,j+1/2,k) = H_z^{2,n-1/2}(i+1/2,j+1/2,k) \\ -\frac{\Delta t}{\mu_z}\left[\frac{\Delta x_2 E_x^{2,n}(i+1/2,j,k)}{\Delta x_0 \Delta y_0} + \frac{\Delta y_2 E_y^{2,n}(i,j+1/2,k)}{\Delta x_0 \Delta y_0}\right] \quad (3.3)$$

Unlike H_z^1 and H_z^2, the updating of H_z requires the electric fields residing on the edges at both sides, though the cell associated with H_z is not split. Specifically, the magnetic field H_z can be updated by using the formula given here:

$$H_z^{n+1/2}(i+3/2,j+1/2,k) = H_z^{n-1/2}(i+3/2,j+1/2,k) \\ +\frac{\Delta t}{\mu_z}\left[\begin{array}{c}\dfrac{\Delta x_0 E_x^n(i+3/2,j+1,k) - \Delta x_0 E_x^n(i+3/2,j,k)}{\Delta x_0 \Delta y_0} \\ -\dfrac{\Delta y_0 E_y^n(i+2,j+1/2,k) - \Delta y_1 E_y^{1,n}(i+1,j+1/2,k) - \Delta y_1 E_y^{2,n}(i+1,j+1/2,k)}{\Delta x_0 \Delta y_0}\end{array}\right] \quad (3.4)$$

Finally the update equations for the truncated electric fields E_y^1 and E_y^2 can be computed via the following expressions:

$$E_y^{1,n+1}(i,j+1/2,k) = \frac{\varepsilon_{1,y} - 0.5\sigma_{1,y}\Delta t}{\varepsilon_{1,y} + 0.5\sigma_{1,y}\Delta t} E_y^{1,n}(i,j+1/2,k) \\ +\frac{\Delta t}{\varepsilon_{1,y} + 0.5\sigma_{1,y}\Delta t}\left[\begin{array}{c}\dfrac{H_x^{n+1/2}(i,j+1/2,k+1/2) - H_x^{n+1/2}(i,j+1/2,k-1/2)}{\Delta z_0} \\ -\dfrac{H_z^{n+1/2}(i+1/2,j+1/2,k) - H_z^{1,n+1/2}(i-1/2,j+1/2,k)}{\Delta x_0}\end{array}\right] \quad (3.5a)$$

$$E_y^{2,n+1}(i,j+1/2,k) = \frac{\varepsilon_{2,y} - 0.5\sigma_{2,y}\Delta t}{\varepsilon_y^2 + 0.5\sigma_y^2\Delta t} E_y^{2,n}(i,j+1/2,k) \\ +\frac{\Delta t}{\varepsilon_{2,y} + 0.5\sigma_{2,y}\Delta t}\left[\begin{array}{c}\dfrac{H_x^{n+1/2}(i,j+1/2,k+1/2) - H_x^{n+1/2}(i,j+1/2,k-1/2)}{\Delta z_0} \\ -\dfrac{H_z^{n+1/2}(i+1/2,j+1/2,k) - H_z^{2,n+1/2}(i-1/2,j+1/2,k)}{\Delta x_0}\end{array}\right] \quad (3.5b)$$

3.2 CONFORMAL TECHNIQUE FOR DIELECTRIC

The conformal technique used for dielectrics is different from that for PEC objects, but also plays a very important role in an efficient FDTD simulation. There exist two major differences between the PEC and dielectric versions of conformal technique: first, in contrast to the situation for a closed PEC object, the fields at both sides of the dielectric interface are nonzero; second, even when the dielectric interface coincides with the FDTD grid, we still need to process the material information on the interface. This is because the FDTD algorithm is designed to solve a differential equation, and the boundary conditions are not explicitly included in the update equations. In view of these differences, the conformal techniques for dielectrics are sometimes referred to as those utilizing effective dielectric boundaries.

To describe the conformal dielectric method, we begin with the situation where the dielectric interface overlaps with the FDTD grid plane, as shown in Figure 3.8. For the tangential electric field components on the boundary, the magnetic field integration loops are now separated into two parts, and they are embedded in different dielectrics. When the two materials are dissimilar as the loop traverses across the interface, the electric field cannot be simply taken as the electric flux divided by the cell size, because this is only valid in homogeneous media.

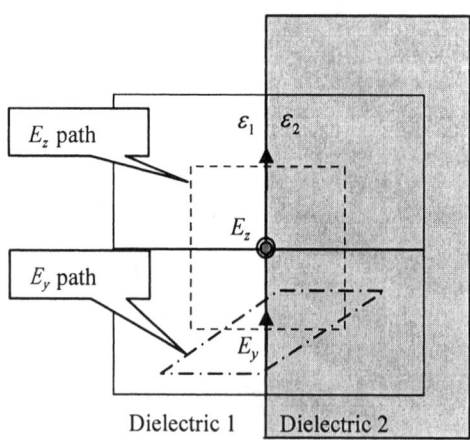

Figure 3.8 Conformal technique when the dielectric interface overlaps with the FDTD grid.

An effective way to deal with this situation is to take the weighted average of the two dielectric constants and then assign it to the cell that contains the magnetic field loop. Simple as it is, this method produces satisfactory results in most applications. For example, in calculating the surface impedance of a dielectric, the solution obtained using this approach agrees very well with the analytical result.

Beginning in the 1990s, there have been several developments of techniques to deal with curved dielectric interfaces. In most of these schemes [5–9], the effective dielectric constant is calculated on the basis of a partially filled volume in a 3-D cell. These approaches, though they improve upon the staircasing approximation, are quite involved in their implementation. In addition, sometimes they introduce inconsistencies, as for instance in the situation depicted in Figure 3.9. In this example, it is evident that the electric field E_y resides in the homogeneous medium-2. However, when we fill a cell with a new material whose dielectric constant is equal to the weighted average of medium-1 and medium-2, we inadvertently locate E_y on a dielectric interface, which is not the physical situation.

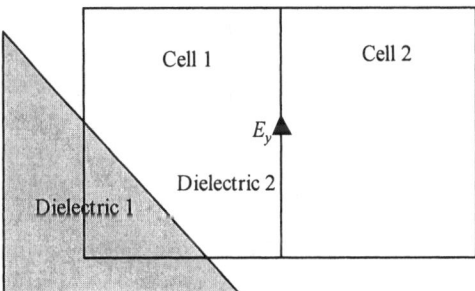

Figure 3.9 Conformal technique for curved dielectric interfaces.

To avoid this difficulty, a slightly different approach has been proposed in [9], in which we do not calculate the effective dielectric constant using 2-D or 3-D material distribution. Instead, we find the intersections of the interface and the grid lines, and if a grid line of a cell is cut into two sections by the dielectric interfaces, we assign an average value of the dielectric constant, determined by weighting the material parameters by the lengths of the grid line in each medium, as illustrated in Figure 3.10.

Let us suppose the medium-1 and medium-2 have the material properties ε_1, σ_1 and ε_2, σ_2, respectively, and Δx_1 and Δx_2 are the lengths of the grid line inside the two media. The effective dielectric constant for this grid line is then:

$$\varepsilon_x^{eff} = \frac{\varepsilon_2 (\Delta x - \Delta x_1) + \varepsilon_1 \Delta x_1}{\Delta x} \qquad (3.6)$$

$$\varepsilon_y^{\text{eff}} = \frac{\varepsilon_2\left(\Delta y - \Delta y_1\right) + \varepsilon_1 \Delta y_1}{\Delta y} \qquad (3.7)$$

$$\sigma_x^{\text{eff}} = \frac{\sigma_2\left(\Delta x - \Delta x_1\right) + \sigma_1 \Delta x_1}{\Delta x} \qquad (3.8)$$

$$\sigma_y^{\text{eff}} = \frac{\sigma_2\left(\Delta y - \Delta y_1\right) + \sigma_1 \Delta y_1}{\Delta y} \qquad (3.9)$$

where $\Delta x_1 = \Delta x - \Delta x_2$ and $\Delta y_1 = \Delta y - \Delta y_2$. From the above equations, we can see that the dielectric constant assigned to a grid line intersected by the interface is unique and is not ambiguous as in the previous scheme. Thus, this scheme for conformal dielectrics is the preferred one that we use in the parallel FDTD algorithm.

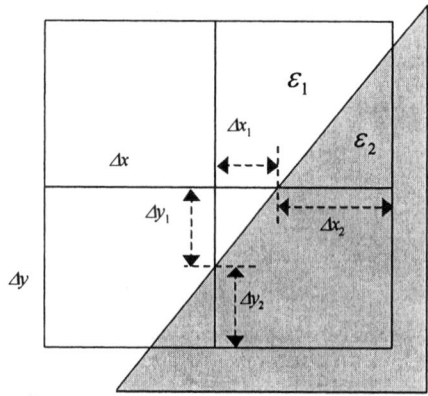

Figure 3.10 Improved conformal technique for curved dielectric interfaces.

3.3 ALTERNATING DIRECTION IMPLICIT (ADI) ALGORITHM

We now turn to a topic that pertains to a scheme designed to handle a fine-grid region embedded within a coarse-grid FDTD mesh. One approach to dealing with the fine-grid region without decreasing the time step in accordance with the Courant condition is to use the backward differencing scheme, which is unconditionally stable. The caveat is that the update equations in this scheme are no longer explicit, and consequently, we must solve a set of linear equations at each time step. For a 3-D problem, solving such an equation is very time- and memory-consuming, even though the associated matrix may be relatively sparse.

Thus the differencing scheme is rarely, if ever, used in practice. To circumvent this difficulty, Namiki and Ito [10] proposed a new scheme called the alternative direction implicit (ADI) algorithm, in which the equations are implicit only in one direction, while the iteration is still unconditionally stable [10]. The above algorithm was later improved by Zheng et al. [11] and has recently been widely applied to many scientific researches and engineering problems. In this section, we introduce the ADI-FDTD algorithm and also discuss the related topic of PML implementation for this method.

3.3.1 Update for ADI Equations

The ADI-FDTD algorithm divides the FDTD procedure at each time step into two steps, each of which utilizes a time step of $\Delta t/2$. In the first half-step, the derivative of E_x is approximated using the backward difference at the time instant $t = (n+1/2)\Delta t$, while the spatial derivatives of the two related magnetic field components H_z and H_y are approximated by their central differences at $y = j\Delta y$ and $z = k\Delta z$, respectively. Unlike the conventional FDTD, the field component H_z is sampled at the time instants $t = (n+1/2)\Delta t$, but H_y is sampled at $t = n\Delta t$. In the update equations, the values of E_x and H_y are known at $t = n\Delta t$, while those for E_x and H_z are unknown at $t = (n+1/2)\Delta t$. Since H_z varies along the y-direction, the number of H_z unknowns is ny (ranging 0 to $ny-1$). On the domain boundary, the value of H_z is obtained from the boundary condition at the time instant $t = (n+1/2)\Delta t$. At the same time instant, the field component E_x is also unknown, and a set of equations can be derived for E_x or H_z by substituting the set of equations into the corresponding H_z, for instance. Thus, once one of them has been determined, the other one can be obtained from an explicit relationship. Similar relationships exist between the other field components, E_y and H_x and E_z and H_y. The update equations are tabulated in (3.10) through (3.15), for completeness.

$$\frac{E_x^{n+1/2}(i+1/2,j,k) - E_x^n(i+1/2,j,k)}{\Delta t/2} = \frac{1}{\varepsilon_x}\left[\begin{array}{c} \dfrac{H_z^{n+1/2}(i+1/2,j+1/2,k) - H_z^{n+1/2}(i+1/2,j-1/2,k)}{\Delta y} \\ -\dfrac{H_y^n(i+1/2,j,k+1/2) - H_y^n(i+1/2,j,k-1/2)}{\Delta z} \end{array}\right] \quad (3.10)$$

$$\frac{E_y^{n+1/2}(i,j+1/2,k) - E_y^n(i,j+1/2,k)}{\Delta t/2} =$$

$$\frac{1}{\varepsilon_y}\left[\begin{array}{c} \dfrac{H_x^{n+1/2}(i,j+1/2,k+1/2) - H_x^{n+1/2}(i,j-1/2,k-1/2)}{\Delta z} \\ -\dfrac{H_z^n(i+1/2,j+1/2,k) - H_z^n(i-1/2,j+1/2,k)}{\Delta x} \end{array}\right] \quad (3.11)$$

$$\frac{E_z^{n+1/2}(i,j,k+1/2) - E_z^n(i,j,k+1/2)}{\Delta t/2} =$$

$$\frac{1}{\varepsilon_z}\left[\begin{array}{c} \dfrac{H_y^{n+1/2}(i+1/2,j,k+1/2) - H_y^{n+1/2}(i-1/2,j,k+1/2)}{\Delta x} \\ -\dfrac{H_x^n(i,j+1/2,k+1/2) - H_x^n(i,j-1/2,k+1/2)}{\Delta y} \end{array}\right] \quad (3.12)$$

$$\frac{H_x^{n+1/2}(i,j+1/2,k+1/2) - H_x^n(i,j+1/2,k+1/2)}{\Delta t/2} =$$

$$\frac{1}{\mu_x}\left[\begin{array}{c} \dfrac{E_y^{n+1/2}(i,j+1/2,k+1) - E_y^{n+1/2}(i,j+1/2,k)}{\Delta z} \\ -\dfrac{E_z^n(i,j+1,k+1/2) - E_z^n(i,j,k+1/2)}{\Delta y} \end{array}\right] \quad (3.13)$$

$$\frac{H_y^{n+1/2}(i+1/2,j,k+1/2) - H_y^n(i+1/2,j,k+1/2)}{\Delta t/2} =$$

$$\frac{1}{\mu_y}\left[\begin{array}{c} \dfrac{E_z^{n+1/2}(i+1,j,k+1/2) - E_z^{n+1/2}(i,j,k+1/2)}{\Delta x} \\ -\dfrac{E_x^n(i+1/2,j,k+1) - E_x^n(i+1/2,j,k)}{\Delta z} \end{array}\right] \quad (3.14)$$

$$\frac{H_z^{n+1/2}(i+1/2,j+1/2,k) - H_z^n(i+1/2,j+1/2,k)}{\Delta t/2} =$$

$$\frac{1}{\mu_z}\left[\begin{array}{c} \dfrac{E_x^{n+1/2}(i+1/2,j+1,k) - E_x^{n+1/2}(i+1/2,j-1,k)}{\Delta y} \\ -\dfrac{E_y^n(i+1,j+1/2,k) - E_y^n(i,j+1/2,k)}{\Delta x} \end{array}\right] \quad (3.15)$$

We note that both (3.10) and (3.15) contain the unknowns E_x and H_z at $t = (n+1/2)\Delta t$; hence, we can obtain an equation just for E_x only by substituting (3.15) into (3.10). We also note that in each equation, three E_x components located at different y positions are involved, and these form a matrix equation for every j. Since only the three neighboring unknowns are involved in each equation, the matrix is tridiagonal. Solving for such a matrix equation is relatively straightforward, but this has to be done at each time step. Similar equations need to be solved for the rest of the field components in the x-z and y-z planes.

Once the field E_x at $t = (n+1/2)\Delta t$ has been determined, we can substitute it into (3.15) to solve for the magnetic field H_z at the same time instant. Similarly, we can use (3.11) and (3.13) to solve for E_y and H_x at $t = (n+1/2)\Delta t$, and use (3.12) and (3.14) to solve for E_z and H_y at $t = (n+1/2)\Delta t$. Following a similar procedure, we can compute the fields at $t = (n+1)\Delta t$ in the second half-time step. The relevant field update equations are given by:

$$\frac{E_x^{n+1}(i+1/2,j,k) - E_x^{n+1/2}(i+1/2,j,k)}{\Delta t/2} = \frac{1}{\varepsilon_x} \left[\begin{array}{c} \dfrac{H_z^{n+1/2}(i+1/2,j+1/2,k) - H_z^{n+1/2}(i+1/2,j-1/2,k)}{\Delta y} \\ -\dfrac{H_y^{n+1}(i+1/2,j,k+1/2) - H_y^{n+1}(i+1/2,j,k-1/2)}{\Delta z} \end{array} \right] \quad (3.16)$$

$$\frac{E_y^{n+1}(i,j+1/2,k) - E_y^{n+1/2}(i,j+1/2,k)}{\Delta t/2} = \frac{1}{\varepsilon_y} \left[\begin{array}{c} \dfrac{H_x^{n+1/2}(i,j+1/2,k+1/2) - H_x^{n+1/2}(i,j-1/2,k-1/2)}{\Delta z} \\ -\dfrac{H_z^{n+1}(i+1/2,j+1/2,k) - H_z^{n+1}(i-1/2,j+1/2,k)}{\Delta x} \end{array} \right] \quad (3.17)$$

$$\frac{E_z^{n+1}(i,j,k+1/2) - E_z^{n+1/2}(i,j,k+1/2)}{\Delta t/2} = \frac{1}{\varepsilon_z} \left[\begin{array}{c} \dfrac{H_y^{n+1/2}(i+1/2,j,k+1/2) - H_y^{n+1/2}(i-1/2,j,k+1/2)}{\Delta x} \\ -\dfrac{H_x^{n+1}(i,j+1/2,k+1/2) - H_x^{n+1}(i,j-1/2,k+1/2)}{\Delta y} \end{array} \right] \quad (3.18)$$

$$\frac{H_x^{n+1}(i,j+1/2,k+1/2) - H_x^{n+1/2}(i,j+1/2,k+1/2)}{\Delta t/2} =$$
$$\frac{1}{\mu_x}\left[\frac{E_y^{n+1/2}(i,j+1/2,k+1) - E_y^{n+1/2}(i,j+1/2,k)}{\Delta z} - \frac{E_z^{n+1}(i,j+1,k+1/2) - E_z^{n+1}(i,j,k+1/2)}{\Delta y}\right] \quad (3.19)$$

$$\frac{H_y^{n+1}(i+1/2,j,k+1/2) - H_y^{n+1/2}(i+1/2,j,k+1/2)}{\Delta t/2} =$$
$$\frac{1}{\mu_y}\left[\frac{E_z^{n+1/2}(i+1,j,k+1/2) - E_z^{n+1/2}(i,j,k+1/2)}{\Delta x} - \frac{E_x^{n+1}(i+1/2,j,k+1) - E_x^{n+1}(i+1/2,j,k)}{\Delta z}\right] \quad (3.20)$$

$$\frac{H_z^{n+1}(i+1/2,j+1/2,k) - H_z^{n+1/2}(i+1/2,j+1/2,k)}{\Delta t/2} =$$
$$\frac{1}{\mu_z}\left[\frac{E_x^{n+1/2}(i+1/2,j+1,k) - E_x^{n+1/2}(i+1/2,j-1,k)}{\Delta y} - \frac{E_y^{n+1}(i+1,j+1/2,k) - E_y^{n+1}(i,j+1/2,k)}{\Delta x}\right] \quad (3.21)$$

In (3.16), E_x and H_y are both sampled at $t = (n+1)\Delta t$; hence, the update equation for E_x is not explicit. Equation (3.20) also contains the same set of unknowns at the same time instant, and therefore we can combine them to solve for E_x and H_y. If we substitute (3.20) into (3.16), we would obtain a set of equations for E_x only. In each equation, three E_x components at different z-positions are involved. The solution process in the second half-time step is the same as that in the first half, except that the directions of the implicit equations are different. We reduce the original 3-D implicit update equation to many 1-D problems. Both the computer memory and simulation time are greatly reduced in the ADI-FDTD method in comparison to that needed to solve 3-D implicit equations. This feature is highly desirable from computational point of view.

The two steps described above complete one iteration in the ADI-FDTD algorithm, and are applied at every time step in the FDTD simulation. In contrast to the conventional FDTD, the ADI-FDTD algorithm is unconditionally stable; thus, the solution will always converge no matter how large a time step we choose in the ADI algorithm. Of course, a solution that converges is not necessarily accurate. For the solution to be accurate, the time step must still be limited by the cell size. The reader is encouraged to find more details on this topic in the literature [11], and we simply present some guidelines below.

Suppose the time step in the conventional FDTD is Δt_{FDTD} and it satisfies the Courant condition (1.23); also let the time step in the ADI-FDTD algorithm be $\Delta t_{\text{ADI}} = \alpha \Delta t_{\text{FDTD}}$, where α is the time factor in the ADI-FDTD algorithm. In order to achieve an accurate solution, the time step should satisfy [12] the condition:

$$\lambda_{\min} / \left(\alpha \max \left(\Delta x, \Delta y, \Delta z \right) \right) \geq 15 \tag{3.22}$$

The application of the ADI-FDTD is not as straightforward as it may appear at first sight, since we have to solve a set of linear equations at each half-time step. Typically the ADI-FDTD iteration is 6–8 times longer than the conventional FDTD update. From (3.22) we can see that if we set α to be 8, then the number of cells in one wavelength should be at least 120 for the ADI-FDTD algorithm. In practice, the use of such a dense mesh is not very common except in circuit boards, packages, or IC chips. Such problems are usually too large to be solved with a single computer and one resorts to parallel processing to simulate such problems. However, the parallel ADI-FDTD algorithm requires a sharing of information between the subdomains, and this makes the efficiency of this algorithm considerably lower than that of the conventional FDTD.

3.3.2 PML Truncation Techniques for ADI-FDTD

In this section, we introduce the time-convolution PML formulation for the ADI-FDTD [12]. The time-convolution PML has been successfully applied to truncate the FDTD computational domain and its superior performance at the low frequencies has been well documented. In the ADI-FDTD algorithm, the update equations are different from those in the conventional FDTD, and hence, we need to rederive the PML formulations for this algorithm. The original split-field PML and Gedney's unsplit field PML all pertain to the update equations, and the anisotropic materials they use in the PML region make the ADI-FDTD iteration more complex. The time-convolution PML, on the other hand, is based on the stretched-coordinates PML formulation, and is not related to the material inside the PML region; hence, it is easier to implement. In the time convolution PML formulation, the update equation is the same as those in the normal update equation, except for the presence of two extra terms at the right-hand side that can be obtained from explicit iterations.

For example, for the E_x component, the update equation in the PML region can be written as [12]:

$$E_x^{n+1/2}(i+1/2,j,k) = E_x^n(i+1/2,j,k)$$
$$+ \frac{\Delta t}{2\varepsilon_0 K_y(j)\Delta y}\left[H_z^{n+1/2}(i+1/2,j+1/2,k) - H_z^{n+1/2}(i+1/2,j-1/2,k)\right]$$
$$- \frac{\Delta t}{2\varepsilon_0 K_z(k)\Delta z}\left[H_y^n(i+1/2,j,k+1/2) - H_y^n(i+1/2,j,k-1/2)\right] \quad (3.23)$$
$$+ \Psi_{exy}^n(i+1/2,j,k) - \Psi_{exz}^n(i+1/2,j,k)$$

where Ψ_{exy}^n and Ψ_{exz}^n can be expressed as:

$$\Psi_{exy}^n(i+1/2,j,k) = b_z(k)\Psi_{exy}^{n-1/2}(i+1/2,j,k)$$
$$+ a_z(k)\frac{H_z^n(i+1/2,j+1/2,k) - H_z^n(i+1/2,j-1/2,k)}{\Delta y} \quad (3.24)$$

$$\Psi_{exz}^n(i+1/2,j,k) = b_y(j)\Psi_{exz}^{n-1/2}(i+1/2,j,k)$$
$$+ a_y(j)\frac{H_y^n(i+1/2,j,k+1/2) - H_y^n(i+1/2,j,k-1/2)}{\Delta z} \quad (3.25)$$

and the coefficients a and b are given by:

$$b_s = \exp\left[-\left(\sigma_{s,\text{PML}}/K_s + \alpha_s\right)\Delta t/2\varepsilon_0\right] \quad (3.26)$$

$$a_s = \frac{\sigma_{s,\text{PML}}}{K_s(\sigma_s + K_s\alpha_s)}(b_s - 1), \quad s = x,y,z \quad (3.27)$$

Similarly, the update equation for H_z is:

$$H_z^{n+1/2}(i+1/2, j+1/2, k) = H_z^n(i+1/2, j+1/2, k)$$
$$+ \frac{\Delta t}{2\mu_0 K_y(i+1/2)\Delta y}\left[E_x^{n+1/2}(i+1/2, j+1, k) - E_x^{n+1/2}(i+1/2, j, k)\right]$$
$$- \frac{\Delta t}{2\mu_0 K_x(i+1/2)\Delta x}\left[E_y^n(i+1, j+1/2, k+1/2) - E_y^n(i, j+1/2, k)\right] \quad (3.28)$$
$$+ \Psi_{hzy}^n(i+1/2, j+1/2, k) - \Psi_{hzx}^n(i+1/2, j+1/2, k)$$

in which Ψ_{hzy} and Ψ_{hzx} have the same forms as those in (3.24) and (3.25). Substituting (3.28) into (3.23), we obtain a set of linear equations to be solved for $E_x^{n+1/2}$ implicitly, and then $H_z^{n+1/2}$ can be explicitly calculated by using (3.28). A similar procedure can be followed in the second step of the ADI-FDTD algorithm to compute the field values at $t = (n+1)\Delta t$. From the above process, we can see that the time-convolution PML is applied by adding two extra terms in the update equations, but with little increase of the complexity of the ADI-FDTD algorithm. Another important feature of the implementation is that, in common with the PML employed in the conventional FDTD, it is not affected by the materials inside the computational region.

3.3.3 Separable Backward-Central Difference

In some specific applications, such as patch antenna problems, the cell size in the vertical direction is much smaller than that in the other two horizontal directions. Because the time step in the FDTD simulations is dominated by the smallest cell size, the small discretization in one direction becomes the bottleneck in problems of this type. Although the separable backward-central difference implicit technique is still limited by the Courant condition, its time step is no longer limited by the smallest cell size as it is in the conventional central difference. We will now demonstrate the application of the back-central difference method by using a 2-D example.

For a TE wave propagation in the z-direction, the three scalar differential equations are:

$$\varepsilon \frac{\partial E_x}{\partial t} = \frac{\partial H_z}{\partial y} \quad (3.29a)$$

$$\varepsilon \frac{\partial E_y}{\partial t} = -\frac{\partial H_z}{\partial x} \qquad (3.29b)$$

$$\mu \frac{\partial H_z}{\partial t} = \frac{\partial E_x}{\partial y} - \frac{\partial E_y}{\partial x} \qquad (3.29c)$$

Let us suppose, for the sake of illustration, that the cell size Δy is much smaller than Δx, and we approximate (3.29a) and (3.29c) using backward differences, while discretizing (3.29b) using central differences. In order to ensure that the solution is stable, E_x and H_z appearing at the right-hand side of (3.29a) and (3.29c) are approximated at $(n+1)\Delta t$ by the average of their values at $(n+1)\Delta t$ and $n\Delta t$. The finite difference form of (3.29) then becomes:

$$\varepsilon \frac{E_x^{n+1} - E_x^n}{\Delta t} = \frac{1}{2} \frac{\Delta\left(H_z^{n+1} + H_z^n\right)}{\Delta y} \qquad (3.30a)$$

$$\varepsilon \frac{E_y^{n+1/2} - E_y^{n-1/2}}{\Delta t} = -\frac{\Delta H_z^n}{\Delta x} \qquad (3.30b)$$

$$\mu \frac{H_z^{n+1} - H_z^n}{\Delta t} = \frac{1}{2} \frac{\Delta\left(E_x^{n+1} + E_x^n\right)}{\Delta y} - \frac{\Delta E_y^{n+1/2}}{\Delta x} \qquad (3.30c)$$

From (3.30), we can see that the electric field E_x and the magnetic field H_z are sampled at integral time steps, while the electric field E_y is sampled at half-time steps. Since the magnetic field term at the left-hand side of (3.30c) is taken at half-time steps, it is taken as the average of H_z^{n+1} and H_z^n. Equations (3.30a) to (3.30c) can then be written as:

$$E_x^{n+1}(i+1/2, j) = E_x^n(i+1/2, j)$$
$$+ \frac{\Delta t}{2\varepsilon \Delta y}\left[H_z^{n+1}(i+1/2, j+1/2) - H_z^{n+1}(i+1/2, j-1/2)\right] \qquad (3.31a)$$
$$+ \frac{\Delta t}{2\varepsilon \Delta y}\left[H_z^n(i+1/2, j+1/2) - H_z^n(i+1/2, j-1/2)\right]$$

$$E_y^{n+1/2}(i,j+1/2) = E_y^{n-1/2}(i,j+1/2) - \frac{\Delta t}{\varepsilon \Delta x}\begin{bmatrix} H_z^n(i+1/2,j+1/2) \\ -H_z^n(i-1/2,j+1/2) \end{bmatrix} \quad (3.31b)$$

$$H_z^{n+1}(i+1/2,j+1/2) = H_z^n(i+1/2,j+1/2)$$
$$+\frac{\Delta t}{2\mu\Delta y}\begin{bmatrix} E_x^{n+1}(i+1/2,j+1) - E_x^{n+1}(i+1/2,j) \\ +E_x^n(i+1/2,j+1) - E_x^n(i+1/2,j) \end{bmatrix} \quad (3.31c)$$
$$-\frac{\Delta t}{\mu\Delta x}\left[E_y^{n+1/2}(i+1,j+1/2) - E_y^{n+1/2}(i,j+1/2) \right]$$

Substituting (3.31c) into (3.31a), we get:

$$-\frac{\beta}{4}E_x^{n+1}(i+1/2,j+1) + \left(1+\frac{\beta}{2}\right)E_x^{n+1}(i+1/2,j) - \frac{\beta}{4}E_x^{n+1}(i+1/2,j-1)$$
$$= E_x^n(i+1/2,j) + \frac{\Delta t}{\varepsilon\Delta y}\left[H_z^n(i+1/2,j+1/2) - H_z^n(i+1/2,j-1/2) \right]$$
$$+\frac{\Delta t^2}{4\varepsilon\mu}\begin{bmatrix} \dfrac{E_x^n(i+1/2,j+1) - E_x^n(i+1/2,j)}{\Delta y \Delta y} \\ -\dfrac{E_x^n(i+1/2,j) - E_x^n(i+1/2,j-1)}{\Delta y \Delta y} \end{bmatrix} \quad (3.32)$$
$$-\frac{\Delta t}{\mu}\begin{bmatrix} \dfrac{E_y^{n+1/2}(i+1,j+1/2) - E_y^{n+1/2}(i,j+1/2)}{\Delta x \Delta y} - \\ \dfrac{E_y^{n+1/2}(i+1,j-1/2) - E_y^{n+1/2}(i,j-1/2)}{\Delta x \Delta y} \end{bmatrix}$$

where $\beta = \Delta t^2 / \varepsilon\mu\Delta y\Delta y$. For a nonuniform mesh and/or inhomogeneous medium, the mesh Δy and/or the magnetic parameter μ are a function of index j, and the coefficient of $E_x^{n+1}(i+1/2,j)$ will also be different from that in (3.32). By varying the index j along the y-direction, and combining (3.31a) and (3.31c), we can obtain a set of linear equations for E_x^{n+1}. The solution of these equations leads us to E_x^{n+1}, from which we can derive H_z^{n+1} explicitly using (3.31c). The electric field $E_y^{n+1/2}$ is explicitly updated next from (3.31b).

Compared with the conventional FDTD, both the backward and central difference schemes are used in this algorithm, and numerical experiments show that it is more accurate than the ADI-FDTD when employing the same time step. In this back-central difference approach, the time step is determined by the Courant condition applied to the cell size in the x-direction, because the field is explicitly updated in this direction.

The stability condition for the above scheme is derived as follows. For a TE wave, the fields can be expressed as:

$$\tilde{E}_x = E_{x0} e^{j\omega t - j(k_x x + k_y y)} \tag{3.33a}$$

$$\tilde{E}_y = E_{y0} e^{j\omega t - j(k_x x + k_y y)} \tag{3.33b}$$

$$\tilde{H}_z = H_{z0} e^{j\omega t - j(k_x x + k_y y)} \tag{3.33c}$$

where the factor $e^{-jk_x x}$ in (3.33) implies a wave propagating in the $+x$-direction, while $e^{jk_x x}$ is associated with a wave propagating in the $-x$-direction. Substituting (3.33) into (3.30), we get:

$$\varepsilon \sin\left(\frac{\omega \Delta t}{2}\right) E_{x0} = \frac{\Delta t}{\Delta y} H_{z0} \cos\left(\frac{\omega \Delta t}{2}\right) \sin\left(\frac{k_y \Delta y}{2}\right) \tag{3.34a}$$

$$\varepsilon \sin\left(\frac{\omega \Delta t}{2}\right) E_{y0} = -\frac{\Delta t}{\Delta x} H_{z0} \sin\left(\frac{k_x \Delta x}{2}\right) \tag{3.34b}$$

$$\frac{H_{z0}}{\Delta t} \mu \sin\left(\frac{\omega \Delta t}{2}\right) = \frac{E_{x0}}{\Delta y} \cos\left(\frac{\omega \Delta t}{2}\right) \sin\left(\frac{k_y \Delta y}{2}\right) - \frac{E_{y0}}{\Delta x} \sin\left(\frac{k_x \Delta x}{2}\right) \tag{3.34c}$$

By substituting (3.34a) and (3.34b) into (3.34c), we can obtain:

$$1 - \cos^2\left(\frac{\omega \Delta t}{2}\right) = \left(\frac{c \Delta t}{\Delta y}\right)^2 \cos^2\left(\frac{\omega \Delta t}{2}\right) \sin^2\left(\frac{k_y \Delta y}{2}\right) + \left(\frac{c \Delta t}{\Delta x}\right)^2 \sin^2\left(\frac{k_x \Delta x}{2}\right) \tag{3.35}$$

where $c = \sqrt{1/\varepsilon\mu}$ is the speed of light. Rearranging the above equation, we can write the stability condition as:

$$\cos\left(\frac{\omega \Delta t}{2}\right) = \sqrt{\frac{1 - \left(\frac{\Delta t}{\Delta x}\right)^2 c^2 \sin^2\left(\frac{k_x \Delta x}{2}\right)}{1 + \left(\frac{\Delta t}{\Delta y}\right)^2 c^2 \sin^2\left(\frac{k_y \Delta y}{2}\right)}} \qquad (3.36)$$

From (3.36) we can see that if the rational function inside the square root is positive, the iteration is stable. Note that this condition is only affected by Δx and not Δy.

3.4 SIMULATION OF DISPERSIVE MEDIA

Dispersive media are those whose dielectric constants are frequency dependent. In this section, we discuss several commonly encountered dispersive media, namely, cold plasma, Debye, Lorentz, and Drude materials [13, 14]. Since the dielectric constants for such media vary with frequency, we first ascertain that they can be characterized in the time domain. Otherwise, we would need to run the time domain code one frequency at a time, rather than over a frequency band.

3.4.1 Recursive Convolution

The most commonly used approach to simulating dispersive materials is the recursive convolution method, which can be applied to the cold plasma, as well as Debye and Lorentz materials. Their dispersion characteristics are given by:

Cold plasma: $\quad \varepsilon(\omega) = \varepsilon_0 \left[1 + \dfrac{\omega_p^2}{\omega(j\nu - \omega)} \right] \qquad (3.37)$

Debye material: $\quad \varepsilon(\omega) = \varepsilon_0 \varepsilon_\infty + \varepsilon_0 \sum_{p=1}^{P} \dfrac{\Delta \varepsilon_p}{1 + j\omega \tau_p} \qquad (3.38)$

Lorentz material: $\varepsilon(\omega) = \varepsilon_0 \varepsilon_\infty + \varepsilon_0 \sum_{p=1}^{P} \frac{(\varepsilon_{s,p} - \varepsilon_{\infty,p})\omega_p^2}{\omega_p^2 + 2j\omega\delta_p - \omega^2}$ (3.39)

where $\varepsilon_{s,p}$ is the permittivity at zero frequency, ν is the plasma frequency, $\varepsilon_{\infty,p}$ is the permittivity as the frequency approaches infinity, τ_p is the relaxation time, δ_p is a dummy coefficient, and ω_p is the resonant frequency of the medium. All the above models can be written in a common form as:

$$\varepsilon(\omega) = \varepsilon_0 \varepsilon_\infty + \sum_{p=1}^{P} \frac{A_p}{\omega - W_p}$$ (3.40)

where W_p are complex poles, and A_p are complex coefficients. The cold plasma model has a pole at the origin, while the Lorentz model has conjugate poles. Since these are special cases of (3.40), they do not require special treatment. The reason for us to write the dispersion relationship in this form is that we want to utilize the special properties of the poles to simplify our update equations. The simplification results from the fact that when dealing with dispersive media, the time domain Maxwell's equations now require not only multiplication, but convolution as well. If the permittivity ε in the time domain is expressed as:

$$\varepsilon(t) = \varepsilon_0 \varepsilon_\infty + \varepsilon_0 \chi(t)$$ (3.41)

then Maxwell's curl equation for the magnetic field can be written as:

$$\nabla \times \vec{H} = \frac{\partial \vec{D}}{\partial t} + \sigma \vec{E} = \frac{\partial \left[\varepsilon_0 \varepsilon_\infty \vec{E}(t) + \varepsilon_0 \int_0^t \vec{E}(t-\tau)\chi(\tau)d\tau \right]}{\partial t} + \sigma \vec{E}$$ (3.42)

in which the convolution term can be approximated by:

$$\int_0^t \vec{E}(t-\tau)\chi(\tau)d\tau \approx \sum_{m=0}^{n-1} \vec{E}^{n-m} \int_{m\Delta t}^{(m+1)\Delta t} \chi(\tau)d\tau$$ (3.43)

The finite difference approximation of (3.43) leads to:

$$\nabla \times \vec{H}\Big|^{n+1/2} = \frac{\vec{D}^{n+1} - \vec{D}^n}{\Delta t} + \sigma \frac{\vec{E}^{n+1} - \vec{E}^n}{2}$$ (3.44)

Next, we define $\chi^m = \int_{m\Delta t}^{(m+1)\Delta t} \chi(\tau) d\tau$, and $\Delta\chi^m = \chi^m - \chi^{m+1}$, and express the derivate of $\vec{D}$ in the difference form as:

$$\begin{aligned}
& D_x^{n+1}(i+\tfrac{1}{2},j,k) - D_x^n(i+\tfrac{1}{2},j,k) \\
&= \varepsilon_0\varepsilon_\infty E_x^{n+1} + \varepsilon_0 \sum_{m=0}^{n} E_x^{n+1-m} \chi^m - \varepsilon_0\varepsilon_\infty E_x^n - \varepsilon_0 \sum_{m=0}^{n-1} E_x^{n-m} \chi^m \\
&= \left(\varepsilon_0\varepsilon_\infty + \varepsilon_0\chi^0\right) E_x^{n+1} - \varepsilon_0\varepsilon_\infty E_x^n - \varepsilon_0 \sum_{m=0}^{n-1} E_x^{n-m}\left(\chi^m - \chi^{m+1}\right) \\
&= \left(\varepsilon_0\varepsilon_\infty + \varepsilon_0\chi^0\right) E_x^{n+1} - \varepsilon_0\varepsilon_\infty E_x^n - \varepsilon_0 \sum_{m=0}^{n-1} E_x^{n-m} \Delta\chi^m
\end{aligned} \quad (3.45)$$

Substituting (3.45) into (3.44), we get:

$$\nabla \times \vec{H}\Big|^{n+1/2} = \frac{\left(\varepsilon_0\varepsilon_\infty + \varepsilon_0\chi^0\right)\vec{E}^{n+1} - \varepsilon_0\varepsilon_\infty \vec{E}^n}{\Delta t} - \sigma\frac{\vec{E}^{n+1} - \vec{E}^n}{2} + \frac{\varepsilon_0}{\Delta t}\sum_{m=0}^{n-1} \vec{E}^{n-m}\Delta\chi^m$$

which we can rewrite in a form that closely resembles the FDTD update equation:

$$\nabla \times \vec{H}\Big|^{n+1/2} = \varepsilon_{\it{eff}}\frac{\vec{E}^{n+1} - \vec{E}^n}{\Delta t} + \sigma_{\it{eff}}\frac{\vec{E}^{n+1} - \vec{E}^n}{2} - \frac{\varepsilon_0}{\Delta t}\sum_{m=0}^{n-1} \vec{E}^{n-m}\Delta\chi^m \quad (3.46)$$

where $\varepsilon_{\it{eff}} = \varepsilon_\infty + \frac{\chi^0}{2}$ and $\sigma_{\it{eff}} = \sigma + \frac{\varepsilon_0\chi^0}{\Delta t}$. The update equation for the electric field then becomes:

$$\vec{E}^{n+1} = \frac{\frac{\varepsilon_{\it{eff}}}{\Delta t} - \frac{\sigma_{\it{eff}}}{2}}{\frac{\varepsilon_{\it{eff}}}{\Delta t} + \frac{\sigma_{\it{eff}}}{2}} \vec{E}^n + \frac{1}{\frac{\varepsilon_{\it{eff}}}{\Delta t} + \frac{\sigma_{\it{eff}}}{2}} \nabla \times \vec{H}^{n+1/2} + \frac{\frac{\varepsilon_0}{\Delta t}}{\frac{\varepsilon_{\it{eff}}}{\Delta t} + \frac{\sigma_{\it{eff}}}{2}} \sum_{m=0}^{n-1} \vec{E}^{n-m}\Delta\chi^m \quad (3.47)$$

We can see from (3.47) that when dispersion is involved, we only need to add one additional term in the update equation. But this term is a convolution, whose computation involves the field values from the start of the simulation to the current time step. This is usually very time- and memory-consuming, and sometimes not even realizable. However, for materials that can be described by the model given by (3.40), we can apply the recursive convolution method as explained later. From (3.40), we know that $\chi(\omega) = \sum_{p=1}^{P} \frac{A_p}{\omega - W_p}$. Its time domain counterpart can be written as:

$$\chi(t) = \sum_{p=1}^{P}(jA_p)e^{j(W_p)t}U(t) = \sum_{p=1}^{P}\chi_p(t) \qquad (3.48)$$

For the pth pole, we have:

$$\chi_p^m = (jA_p)\int_{m\Delta t}^{(m+1)\Delta t} e^{jW_p\tau}U(\tau)d\tau = \left(-\frac{A_p}{W_p}\right)e^{jW_p m\Delta t}\left[1-e^{jW_p\Delta t}\right] \qquad (3.49)$$

$$\Delta\chi_p^m = \left(-\frac{A_p}{W_p}\right)e^{jW_p m\Delta t}\left[1-e^{jW_p\Delta t}\right]^2 \qquad (3.50)$$

By defining $\psi_p^n = \sum_{m=0}^{n-1}\vec{E}^{n-m}\Delta\chi_p^m$, the convolution in (3.43) can be written as:

$$\psi_p^n = \vec{E}^n\Delta\chi_p^0 + \sum_{m=1}^{n-1}\vec{E}^{n-m}\Delta\chi_p^m = \vec{E}^n\Delta\chi_p^0 + e^{jW_p\Delta t}\sum_{m=0}^{n-2}\vec{E}^{n-1-m}\Delta\chi_p^m$$
$$= \vec{E}^n\Delta\chi_p^0 + e^{jW_p\Delta t}\psi_p^{n-1} \qquad (3.51)$$

The above equation implies that each of the poles ψ_p^n can be determined from its value at the previous time step, together with the electric field at the current time step. The use of (3.51) circumvents the need to store the field history, and the update equation (3.47) becomes:

$$\vec{E}^{n+1} = \frac{\frac{\varepsilon_{eff}}{\Delta t}-\frac{\sigma_{eff}}{2}}{\frac{\varepsilon_{eff}}{\Delta t}+\frac{\sigma_{eff}}{2}}\vec{E}^n + \frac{1}{\frac{\varepsilon_{eff}}{\Delta t}+\frac{\sigma_{eff}}{2}}\nabla\times\vec{H}^{n+1/2} + \frac{\frac{\varepsilon_0}{\Delta t}}{\frac{\varepsilon_{eff}}{\Delta t}+\frac{\sigma_{eff}}{2}}\sum_{p=1}^{P}\psi_p^n \qquad (3.52)$$

The value of ψ_p^n is updated from the electric field before proceeding to the next step.

In the following example, we compute the scattering field from a dispersive sphere by using the recursive convolution method. The sphere has a radius of 2 mm, and is comprised of a Lorentz material. The relevant parameters in (3.39) are $\omega_p = 40\pi \times 10^9$ GHz, $\varepsilon_{s,p} = 3.0$, $\varepsilon_{\infty,p} = 1.5$, and $\delta_p = 4\pi \times 10^9$. The variation of the dielectric constant versus frequency is plotted in Figure 3.11.

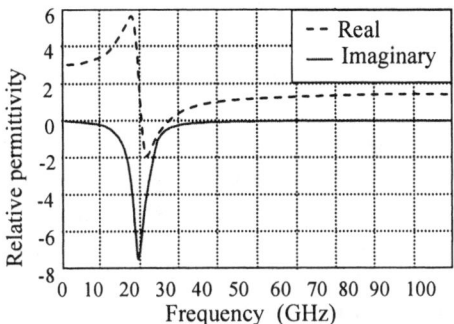

Figure 3.11 Dielectric constant of the dispersive sphere.

The monostatic RCS of the sphere is plotted in Figure 3.12. For the sake of comparison, the analytical result is also plotted in the same figure.

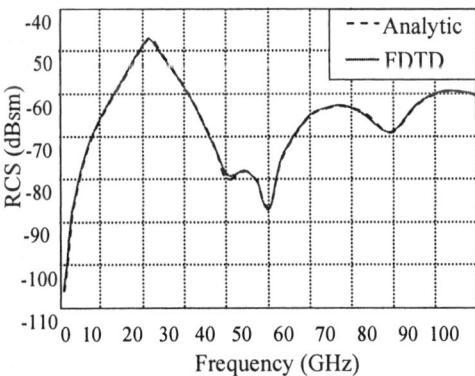

Figure 3.12 Monostatic RCS of the dispersive sphere.

Before ending this section, we need to mention that if the electric field $\vec{E}(t-\tau)$ in (3.43) is a linear relationship with respect to the time t instead of a constant in the time period of $m\Delta t$ to $(m+1)\Delta t$, (3.52) will have a slightly different format and the result will be more accurate [14].

3.4.2 Pole Extraction Techniques

In the previous sections we have dealt with dispersive materials for which physical models are available in the form given in (3.40), which is required for recursive convolution. However, not all materials can be described in the above form; hence, it becomes necessary to approximate the material properties from a pure mathematical point of view, in order to render them amenable to the FDTD simulation.

Assuming that we know the dielectric constants of the material at several frequencies, the problem now becomes finding a model in the form of (3.40) that yields the same dielectric constants at these sample frequencies. Commonly used approaches include Cauchy's method [15] and the Padé approximation technique [16]. Here, we introduce Cauchy's method based on the singular value decomposition technique [17]. The advantage of this approach is that the model order can be estimated. We first write (3.40) into a rational function form:

$$\varepsilon(\omega) = \frac{A(\omega)}{B(\omega)} = \frac{\sum_{p=0}^{p=P-1} a_p \omega^p}{\sum_{p=0}^{p=P} b_p \omega^p} \tag{3.53}$$

Using the prescribed values of $\varepsilon(\omega)$ at, say, N frequency points, we can derive a matrix equation:

$$\begin{bmatrix} \varepsilon(\omega_1)\omega_1^0 & \cdots & \varepsilon(\omega_1)\omega_1^p & \cdots & \varepsilon(\omega_1)\omega_1^P \\ \vdots & \ddots & \vdots & \ddots & \vdots \\ \varepsilon(\omega_n)\omega_n^0 & \cdots & \varepsilon(\omega_n)\omega_n^p & \cdots & \varepsilon(\omega_n)\omega_n^P \\ \vdots & \ddots & \vdots & \ddots & \vdots \\ \varepsilon(\omega_N)\omega_N^0 & \cdots & \varepsilon(\omega_N)\omega_N^p & \cdots & \varepsilon(\omega_N)\omega_N^P \end{bmatrix} \begin{bmatrix} b_0 \\ \vdots \\ b_n \\ \vdots \\ b_N \end{bmatrix} - \begin{bmatrix} \omega_1^0 & \cdots & \omega_1^p & \cdots & \omega_1^P \\ \vdots & \ddots & \vdots & \ddots & \vdots \\ \omega_n^0 & \cdots & \omega_n^p & \cdots & \omega_n^P \\ \vdots & \ddots & \vdots & \ddots & \vdots \\ \omega_N^0 & \cdots & \omega_N^p & \cdots & \omega_N^P \end{bmatrix} \begin{bmatrix} a_0 \\ \vdots \\ a_n \\ \vdots \\ a_N \end{bmatrix} = 0$$

which can be written in a compact form as:

$$[\mathbf{A} \quad -\mathbf{B}] \begin{bmatrix} a \\ b \end{bmatrix} = 0 \tag{3.54}$$

When $N>2P+1$, the matrix $[\mathbf{A} \quad -\mathbf{B}]$ has $2P+1$ singular values. If the actual model order P' is smaller than P, there will be $2P'+1$ singular values that are larger than the others, which correspond to noise. The right singular vector corresponding to the smallest singular value can be regarded as the solution to (3.54) for the coefficients a and b. Once they have been solved, we can find the poles by finding the roots of the polynomial in (3.53), and then derive the coefficients in (3.40) as the residuals of the pole.

3.4.3 Simulation of Double-Negative Materials

In this section, we introduce a commonly used dispersive model for the double-negative materials, namely the Drude model. In this model, the relative permittivity and permeability can be expressed as [18]:

$$\varepsilon(\omega) = 1 - \frac{\omega_p^2}{\omega^2} \tag{3.55}$$

$$\mu(\omega) = 1 - \frac{\omega_p^2}{\omega^2} \tag{3.56}$$

where ω_p is the resonant frequency of the material. The advantage of this model is that it can be used to describe double-negative materials, in which both the permittivity and permeability are negative. From (3.55) and (3.56) we can see that if the resonant frequency ω_p is greater than a certain frequency, both the permittivity and permeability at higher frequency will have a negative value. Such a material is called the left-handed material or double-negative material. Substituting (3.55) and (3.56) into Maxwell's equations, we have:

$$j\omega\mu_0 \tilde{\vec{H}} + \tilde{\vec{K}} = -\nabla \times \tilde{\vec{E}} \tag{3.57}$$

$$j\omega\varepsilon_0 \tilde{\vec{E}} + \tilde{\vec{J}} = \nabla \times \tilde{\vec{H}} \tag{3.58}$$

where $\tilde{\vec{K}} = -j\mu_0 \frac{\omega_p^2}{\omega} \tilde{\vec{H}}$ and $\tilde{\vec{J}} = -j\varepsilon_0 \frac{\omega_p^2}{\omega} \tilde{\vec{E}}$. The corresponding equations in the time domain are then:

$$\mu_0 \frac{\partial \vec{H}}{\partial t} + \vec{K} = -\nabla \times \vec{E} \tag{3.59}$$

$$\varepsilon_0 \frac{\partial \vec{E}}{\partial t} + \vec{J} = \nabla \times \vec{H} \qquad (3.60)$$

$$\frac{\partial \vec{K}}{\partial t} = \mu_0 \omega_p^2 \vec{H} \qquad (3.61)$$

$$\frac{\partial \vec{J}}{\partial t} = \varepsilon_0 \omega_p^2 \vec{E} \qquad (3.62)$$

If we sample $\vec{K}$ and $\vec{E}$ at the same time instants, and $\vec{J}$ and $\vec{H}$ at the same time instants, then the above equations can be written in an explicit form similar to the conventional FDTD equations.

3.5 CIRCUIT ELEMENTS

In this section, we introduce techniques for incorporating commonly used circuit elements, namely resistors, capacitors, and inductors in the FDTD algorithm [19, 20]. In the traditional approaches, the circuit elements are introduced into the FDTD simulation by solving for current that flows through them. The drawback in these approaches is that we have to modify the update equations separately for each type of circuit element, compromising the flexibility of the FDTD algorithm in the process. Here, we introduce a technique in which these circuit elements can be replaced by the corresponding material distribution, obviating the need to modify the update equations. We begin with a resistor whose relationship between the current and voltage can be expressed as $RI = V$ where R is its resistance. We can then write:

$$I^{n+1/2} = \frac{V}{R} = \frac{DE^{n+1} + DE^n}{2R} \qquad (3.63)$$

where D is the length of the resistor. Using the relationship $J = \frac{I}{A}$ where A is the effective area of the inductor, we have:

$$J^{n+1/2} = \frac{DE^{n+1} + DE^n}{2RA} \qquad (3.64)$$

Substituting the current density given by (3.64) into the equation:

$$\nabla \times \vec{H} = \varepsilon \frac{\partial \vec{E}}{\partial t} + \sigma \vec{E} + \vec{J} \qquad (3.65)$$

we obtain the following equation for the *x*-component:

$$\nabla \times \vec{H}\Big|_x^{n+1/2} = \varepsilon_x \frac{E_x^{n+1} - E_x^n}{\Delta t} + \sigma_x \frac{E_x^{n+1} + E_x^n}{2} + J_x^{n+1/2} \qquad (3.66)$$

Substituting the current density in (3.64) into (3.66), we have:

$$\nabla \times \vec{H}\Big|_x^{n+1/2} = \varepsilon_x \frac{E_x^{n+1} - E_x^n}{\Delta t} + \sigma_x \frac{E_x^{n+1} + E_x^n}{2} + \frac{DE_x^{n+1} + DE_x^n}{2RA} \qquad (3.67)$$

We simplify (3.67) to obtain an explicit update formula for E_x^{n+1}, which reads:

$$E_x^{n+1} = \frac{\varepsilon_x - 0.5\left(\sigma_x + \sigma_{x,resistor}^{eff}\right)\Delta t}{\varepsilon_x + 0.5\left(\sigma_x + \sigma_{x,resistor}^{eff}\right)\Delta t} E_x^n + \frac{\Delta t}{\varepsilon_x + 0.5\left(\sigma_x + \sigma_{x,resistor}^{eff}\right)\Delta t} \nabla \times \vec{H}\Big|_x^{n+1/2}$$

$$(3.68)$$

where $\sigma_{x,resistor}^{eff} = D/AR$ is defined as an effective conductivity. The effective area A is determined by the basis functions employed in the FDTD algorithm. The basis functions of the FDTD are step functions for both the electric and magnetic fields, which stems from the fact that these fields are sampled uniformly within the FDTD cell, as shown in Figure 3.13.

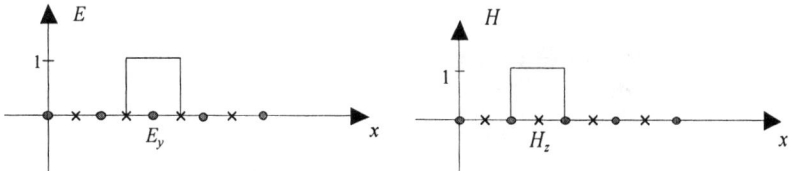

Figure 3.13 Basis functions of the electric and magnetic fields.

In a 2-D space, the magnetic field at a point is represented by an average of the fields in the interior region of the FDTD cells, while the zone for the electric field straddles between four adjacent FDTD cells, as shown in Figure 3.14. Unlike the regular material distribution, the electric fields are never located on the boundary of the effective conductivity corresponding to a resistor, hence, we do not need to apply any specific condition on this boundary.

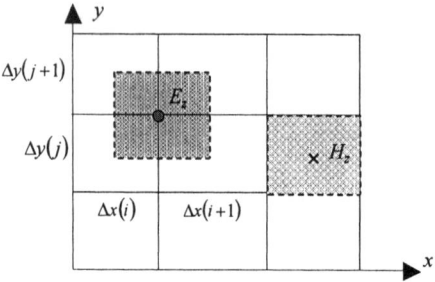

Figure 3.14 The 2-D basis functions of the electric and magnetic fields.

For a given current density J_z, the current flowing through the sampling region can be expressed as:

$$I_z = \int_{\Delta s} J_z \cdot ds = \int_{\Delta s} \sigma E_z \cdot ds$$
$$= \frac{\Delta x(i) + \Delta x(i+1)}{2} \frac{\Delta y(j) + \Delta y(j+1)}{2} \sigma E_z \quad (3.69)$$

Next, turn to the simulation of a capacitor (see Figure 3.15(a)) in the FDTD algorithm, whose equivalent circuit is shown in Figure 3.15(b).

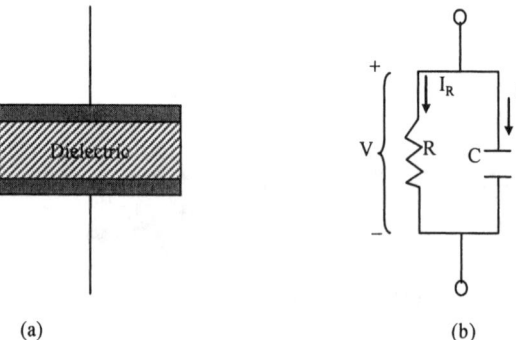

Figure 3.15 Capacitor and its equivalent circuit. (a) Capacitor. (b) Equivalent circuit of capacitor.

Using the relationship $C\,dV/dt = I + I_R$, we can derive the following expression for the discretized representation of the current that flows through the capacitor:

$$I^{n+1/2}(i,j,k) = C\frac{DE^{n+1} - DE^n}{\Delta t} - \frac{DE^{n+1} + DE^n}{2R} \qquad (3.70)$$

where C and D are the capacitance and its length, respectively. Using the relation between the current and the current density, the latter can be expressed as:

$$J^{n+1/2} = C\frac{DE^{n+1} - DE^n}{A\Delta t} - \frac{DE^{n+1} + DE^n}{2AR} \qquad (3.71)$$

Substituting the above expression for the current density into (3.65), we have the following equation for the x-component of the curl equation:

$$\nabla \times \vec{H}\big|_x^{n+1/2} = \varepsilon_x \frac{E_x^{n+1} - E_x^n}{\Delta t} + \sigma_x \frac{E_x^{n+1} + E_x^n}{2} \qquad (3.72)$$
$$+ C\frac{DE_x^{n+1} - DE_x^n}{A\Delta t} - \frac{DE_x^{n+1} + DE_x^n}{2RA}$$

Simplifying (3.72), the update formulation of E_x^{n+1} can be expressed as:

$$E_x^{n+1} = \frac{\varepsilon_x + \varepsilon_{x,capacity}^{eff} - 0.5\left(\sigma_x + \sigma_{x,capacity}^{eff}\right)\Delta t}{\varepsilon_x + \varepsilon_{x,capacity}^{eff} + 0.5\left(\sigma_x + \sigma_{x,capacity}^{eff}\right)\Delta t} E_x^n$$
$$+ \frac{\Delta t}{\varepsilon_x + \varepsilon_{x,capacity}^{eff} + 0.5\left(\sigma_x + \sigma_{x,capacity}^{eff}\right)\Delta t} \nabla \times \vec{H}\big|_x^{n+1/2} \qquad (3.73)$$

where $\varepsilon_{x,capacity}^{eff} = \frac{CD}{A}$ and $\sigma_{x,capacity}^{eff} = \frac{D}{RA}$ are referred to as the effective permittivity and conductivity of a capacitor, respectively.

Finally, we address the problem of modeling a lumped element inductor. In the conventional concept, the inductance in the circuit theory corresponds to the permeability of magnetic material in the electromagnetic fields [21]. It is not convenient to implement a similar relationship in the FDTD simulation due to the available space and the magnetic material filling inside the inductor increases the complexity of the FDTD update equations.

Numerical experiments show that if we directly substitute the relationship $L\,dI/dt + RI = V$ into Maxwell's equations in order to simulate an inductor in the FDTD algorithm, it results in late-time instability when the inductance is small [22]. Though both the fully implicit and semi-implicit schemes can be employed to mitigate the instability problem, a fully implicit FDTD scheme is difficult to parallelize. We introduce the semi-implicit scheme shown below and investigate the relationship between the inductance and the material parameters. An inductor and its equivalent circuit are shown in Figure 3.16.

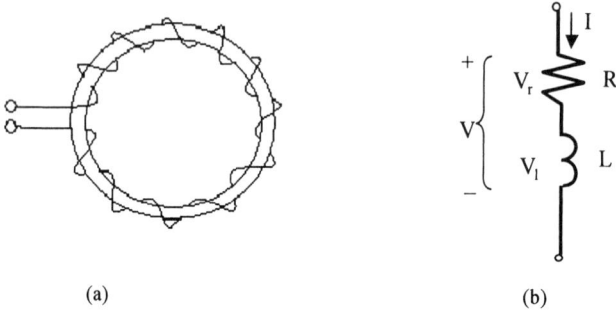

(a)   (b)

Figure 3.16 Inductor and its equivalent circuit. (a) Inductor. (b) Equivalent circuit of inductor.

The discretized relationship between the current density and the fields can be expressed as:

$$LA\frac{J^{n+1}-J^n}{\Delta t} + RA\frac{J^{n+1}+J^n}{2} = D\frac{E^{n+1}+E^n}{2} \quad (3.74)$$

where D and L are the length and inductance value, respectively. The current density can be obtained from the above equation and written in the form:

$$J^{n+1} = J^n + \frac{D\Delta t}{2A}\frac{E^{n+1}+E^n}{\left(L+\frac{R\Delta t}{2}\right)} \quad (3.75)$$

Substituting (3.75) into (3.65), we obtain the following implicit update equation:

$$\nabla\times\vec{H}\Big|_x^{n+1/2} = \varepsilon_x\frac{E_x^{n+1}-E_x^n}{\Delta t} + \sigma_x\frac{E_x^{n+1}+E_x^n}{2} + \frac{J_x^{n+1}+J_x^n}{2} \quad (3.76)$$

Equation (3.76) is implicit because both E_x^{n+1} and J_x^{n+1} are measured at the same time step $(n+1)\Delta t$. In contrast to the conventional implicit schemes, we do not need to solve any matrix equations since J_x^{n+1} appearing in (3.76) can be replaced by the expression given in (3.75) to get:

$$E_x^{n+1} = \frac{\varepsilon_x - 0.5\left(\sigma_x + \sigma_{x,inductor}^{eff}\right)\sigma\Delta t}{\varepsilon_x + 0.5\left(\sigma_x + \sigma_{x,inductor}^{eff}\right)\Delta t} E_x^n + \frac{\Delta t}{\varepsilon_x + 0.5\left(\sigma_x + \sigma_{x,inductor}^{eff}\right)\Delta t} \nabla \times \vec{H}\Big|_x^{n+1/2}$$
$$- \frac{\Delta t}{\varepsilon_x + 0.5\left(\sigma_x + \sigma_{x,inductor}^{eff}\right)\Delta t} J_x^n \tag{3.77}$$

$$\sigma_{x,inductor}^{eff} = \frac{D\Delta t}{2A\left(L + \dfrac{R\Delta t}{2}\right)} \tag{3.78}$$

where $\sigma_{x,inductor}^{eff}$ is referred to as the effective conductance of an inductor. As $L \to 0$, the effective conductance becomes $\sigma_{x,inductor}^{eff} = D/AR$, namely, a resistor. As both the inductance and resistance in the effective conductance are allowed to be tend to 0, $\sigma_{x,inductor}^{eff} \to \infty$ (i.e., we have a case in the limit). It can be verified from (3.77) that in contrast to the conventional approach to modeling a lumped inductance in the FDTD, the update equation (3.77) is always stable regardless of the value of the inductance.

The current density J_x^n in (3.77) can be computed via the following update equation:

$$J_x^n = J_x^{n-1} + \sigma_{x,inductor}^{eff}\left(E_x^n + E_x^{n-1}\right) \tag{3.79}$$

The above update procedure requires two additional arrays for the previous current density and electric fields inside the inductor region. Obviously, an inductor is an element that stores energy rather than one that dissipates energy; hence, a representation of L in terms of an effective conductivity would be counter to the physics. We notice that there is an extra term in (3.77) whose function is just like an excitation source. However, we notice from (3.79) that it contains an excitation term that accounts for the power dissipation in the effective resistance appearing in (3.77). Furthermore, the excitation and resistance work together to mimic the function of an inductor.

In summary, the update equation for a resistor, capacitor, and inductor can be expressed in a general formula as follows:

$$E_x^{n+1} = \frac{\left(\varepsilon_x + \varepsilon_{x,capacitor}^{eff}\right) - 0.5\left(\sigma_x + \sigma_{x,resistor}^{eff} + \sigma_{x,capacitor}^{eff} + \sigma_{x,inductor}^{eff}\right)\sigma\Delta t}{\left(\varepsilon_x + \varepsilon_{x,capacitor}^{eff}\right) + 0.5\left(\sigma_x + \sigma_{x,resistor}^{eff} + \sigma_{x,capacitor}^{eff} + \sigma_{x,inductor}^{eff}\right)\Delta t} E_x^n$$

$$+ \frac{\Delta t}{\left(\varepsilon_x + \varepsilon_{x,capacitor}^{eff}\right) + 0.5\left(\sigma_x + \sigma_{x,resistor}^{eff} + \sigma_{x,capacitor}^{eff} + \sigma_{x,inductor}^{eff}\right)\Delta t} \nabla \times \vec{H}\Big|_x^{n+1/2}$$

$$- \frac{\Delta t}{\left(\varepsilon_x + \varepsilon_{x,capacitor}^{eff}\right) + 0.5\left(\sigma_x + \sigma_{x,resistor}^{eff} + \sigma_{x,capacitor}^{eff} + \sigma_{x,inductor}^{eff}\right)\Delta t} J_x^n$$

(3.80)

with

$$J_x^n = J_x^{n-1} + \sigma_{x,inductor}^{eff}\left(E_x^n + E_x^{n-1}\right) \qquad (3.81)$$

Returning now to the topic of simulation of circuit parameters in general, we note that in simulating the circuit elements, the input parameters chosen in the FDTD simulation are usually the values of the resistors, capacitors, and inductors. If the cross section of circuit element straddles across more than one subdomain in the parallel scheme, we should first compute the equivalent dielectric parameters corresponding to the circuit elements, and then assign the appropriate dielectric parameter distribution to each processor, instead of the values of the resistors, capacitors, or inductors.

Before ending this section, we discuss the simulation of an active element, namely, a lumped diode in the FDTD simulation. The voltage-current relationship for a diode is nonlinear, and can be expressed as [23]:

$$I_d = I_0 \left\{ \exp\left(\frac{qV_d}{nkT}\right) - 1 \right\} \qquad (3.82)$$

where the parameter I_d is the current flowing through the diode and V_d is the voltage measured at its terminals of the diode, respectively. The parameters I_0 and q are the saturation current and the charge of an electron, respectively, while n, k, and T are the emission coefficient, Boltzmann's constant, and temperature, respectively. The FDTD update equation for the diode for the E-field, for instance, is given by:

$$E_z^{n+1} = \frac{\varepsilon_z - 0.5\sigma_z \Delta t}{\varepsilon_z + 0.5\sigma_z \Delta t} E_z^n + \frac{\Delta t}{\varepsilon_z + 0.5\sigma_z \Delta t} \nabla \times \vec{H}\Big|_z^{n+1/2}$$
$$- \frac{\Delta t}{\varepsilon_z + 0.5\sigma_z \Delta t} \frac{I_0}{\Delta s} \left\{ \exp\left[\frac{q\Delta z}{2kT}\left(E_z^{n+1} + E_z^n\right)\right] - 1 \right\} \quad (3.83)$$

which is implicit. If the diode is located within one subdomain, as is very likely to be the case in practice, the performance of the parallel processing algorithm will suffer little because all the other subdomains will remain explicit. Finally, we mention that the diode can also be simulated by modeling its physical geometry [24], though this is not always convenient, because the feature sizes involved are usually very small.

REFERENCES

[1] T. Jurgens, A. Taflove, and K. Moore, "Finite-Difference Time-Domain Modeling of Curved Surfaces [EM Scattering]," *IEEE Transactions on Antennas and Propagation*, Vol. 40, No. 4, April 1992, pp. 357-366.

[2] S. Dey and R. Mittra, "A Locally Conformal Finite-Difference Time-Domain (FDTD) Algorithm for Modeling Three-Dimensional Perfectly Conducting Objects," *IEEE Microwave and Guided Wave Letters*, Vol. 7, No. 9, September 1997, pp. 273-275.

[3] W. Yu and R. Mittra, "A Conformal FDTD Software Package for Modeling of Antennas and Microstrip Circuit Components," *IEEE Antennas and Propagation Magazine*, Vol. 42, No. 5, October 2000, pp. 28-39.

[4] W. Yu and R. Mittra, *Conformal Finite-Difference Time-Domain Maxwell's Equations Solver: Software and User's Guide*, Artech House, Norwood, MA, 2004.

[5] X. Liang and K. Zakim, "Modeling of Cylindrical Dielectric Resonators in Rectangular Waveguides and Cavity," *IEEE Transactions on Microwave Theory and Techniques*, Vol. 41, No. 12, December 1993, pp. 2174-2181.

[6] M. Marcysiak and W. Gwarek, "Higher Order Modeling of Media Surfaces for Enhanced FDTD Analysis of Microwave Circuits," *Proc. 24th European Microwave Conf.*, Vol. 2, Cannes, France, 1994, pp. 1530-1535.

[7] N. Kaneda, B. Houshm, and T. Itoh, "FDTD Analysis of Dielectric Resonators with Curved Surfaces," *IEEE Transactions on Microwave Theory and Techniques*, Vol. 45, No. 9, September 1997, pp. 1645-1649.

[8] S. Dey and R. Mittra, "A Conformal Finite-Difference Time-Domain Technique for Modeling Cylindrical Dielectric Resonators," *IEEE Transactions on Microwave Theory and Techniques*, Vol. 47, No. 9, September 1999, pp. 1737-1739.

[9] W. Yu and R. Mittra, "A Conformal Finite Difference Time Domain Technique for Modeling Curved Dielectric Surfaces," *IEEE Microwave and Wireless Components Letters*, Vol. 11, No. 1, January 2001, pp. 25-27.

[10] T. Namiki, "A New FDTD Algorithm Based on Alternating-Direction Implicit Method," *IEEE Transactions on Microwave Theory and Techniques*, Vol. 47, No. 10, October 1999, pp. 2003-2007.

[11] F. Zheng, Z. Chen, and J. Zhang, "Toward the Development of a Three-Dimensional Unconditionally Stable Finite-Difference Time-Domain Method," *IEEE Transactions on Microwave Theory and Techniques*, Vol. 48, No. 9, September 2000, pp. 1550-1558.

[12] S. Gedney, "Perfectly Matched Layer Media with CFS for an Unconditionally Stable ADI-FDTD Method," *IEEE Transactions on Antennas and Propagation*, Vol. 49, No. 11, November 2001, pp. 1554-1559.

[13] R. Luebbers and F. Hunsberger, "FDTD for Nth-Order Dispersive Media," *IEEE Transactions on Antennas and Propagation*, Vol. 40, No. 11, November 1992, pp. 1297-1301.

[14] J. Young and R. Nelson, "A Summary and Systematic Analysis of FDTD Algorithms for Linearly Dispersive Media," *IEEE Antennas Propagation Magazine*, Vol. 43, No. 1, February 2001, pp. 61-77.

[15] R. Adve, et al., "Application of the Cauchy Method for Extrapolating/Interpolating Narrow-Band System Response," *IEEE Transactions on Microwave Theory and Techniques*, Vol. 45, No. 5, May 1997, pp. 837-845.

[16] C. Brezinski, *Padé-Type Approximation and General Orthogonal Polynomials*, Birkhauser Verlag, Besel, Switzerland, 1980.

[17] K. Kottapalli, T. Sarkar, and Y. Hua, "Accurate Computation of Wide-Band Response of Electromagnetic Systems Utilizing Narrow-Band Information," *IEEE Transactions on Microwave Theory and Techniques*, Vol. 39, No. 4, April 1991, pp. 682-688.

[18] D. Correla and J. Jin, "3D-FDTD-PML Analysis of Left-Handed Metamaterials," *Microwave and Optical Technology Letters*, Vol. 40, No. 3, February 2004, pp. 201-205.

[19] A. Taflove and S. Hagness, *Computational Electromagnetics: The Finite-Difference Time-Domain Method*, Artech House, Norwood, MA, 2000.

[20] M. Piket-May, A. Taflove, and J. Baron, "FD-TD Modeling of Digital Signal Propagation in 3-D Circuits with Passive and Active Loads," *IEEE Transactions on Microwave Theory and Techniques*, Vol. 42, 1994, pp. 1514-1523.

[21] R. Harrington, *Time-Harmonic Electromagnetic Fields*, IEEE Press, Piscataway, NJ, 2001.

[22] W. Thiel and L. Katehi, "Some Aspects of Stability and Numerical Dissipation of the Finite-Difference Time-Domain (FDTD) Techniques Including Passive and Active Lumped Elements," *IEEE Transactions on Microwave Theory and Techniques*, Vol. 5, No. 9, September 2002, pp. 2159-2165.

[23] F. Kung and H. Chuah, "Modeling a Diode in FDTD," *Journal of Electromagnetic Waves and Applications*, Vol. 16, No. 1, 2002, pp. 99-110.

[24] W. Thiel and W. Menzel, "Full-Wave Design and Optimization of mm-Wave Diode-Based Circuits in Finline Technique," *IEEE Transactions on Microwave Theory and Techniques*, Vol. 47, No. 12, December 1999, pp. 2460-2466.

Chapter 4

Excitation Source

In addition to the boundary conditions, the excitation source is also an important component in the FDTD simulation. However, it does not usually draw enough attention since most people only care about the relative result. As a matter of fact, the excitation sources are critical in the FDTD simulation, and its implementation can directly affect the result accuracy and the convergence speed. In this chapter, we introduce the basic concepts and implementations of some fundamental excitation sources in FDTD simulation.

4.1 INTRODUCTION

A source is a mechanism that introduces energy into a system. The actual source can be very complicated and hard to model, and if we model every detail of the source, the discretization can be too small resulting in huge memory consumption and an extremely long simulation time. In such cases, simplification of the model is necessary. For example, many patch antennas are excited by a coax cable. If we model the inner and outer conductor of the coax in our simulation, the mesh grid will be very dense because the inner conductor is very thin. But since we know that the coax is a uniform transmission line, and that the source dimension is much smaller than the antenna, we can readily model the source as a lumped source between the ground and the antenna and place it on one cell. By doing this, small cells are avoided without sacrificing accuracy. From this example, we can see that the source configuration is an important issue in an FDTD simulation.

There are many types of source implementations available. One of them is a so-called "hard" source that forces the fields in the source region to be a predefined time signal. The fields in the source region are not involved in the update procedure until the excitation signal is finished. Its advantage is that the incident wave has the same shape as the excitation signal. Since it causes undesired reflection from the source region, this type of excitation is rarely used in the FDTD simulations today. Another type of excitation is implemented through the source current term in Maxwell's equation. This type of excitation is broadly used in practice because it does not cause reflection from the source region.

The two type of sources mentioned above could be categorized as "local" sources, which are completely contained in the computational domain. For scattering problems, the incident plane wave comes from infinity, and is difficult to implement with local sources. There are two popular ways to solve such problems; the first way is by using the scattered field formulation, which uses the scattered field as unknowns and the incident field as a source term in the update equation. The second method is the total/scattered formulation, in which the unknowns are the total fields inside a subdomain, and the scattered fields elsewhere. In this formulation, the incident field is only applied on the boundaries of the two regions. It is worthwhile mentioning that these two methods can be applied not only to plane wave sources, but also to any source type, including local source.

In addition to the spatial distribution of the source, we also need to be concerned about its time domain variation. For a linear system, the time domain output of the FDTD simulation is the convolution of the system impulse response and the input signal. In the frequency domain, the response is the system frequency response multiplied by the source spectrum. To get the frequency response in a certain frequency band, the source spectrum or its frequency derivative cannot both be zero or very small in the band. On the other hand, some structures in the model may cause a resonance in the simulation, and some resonant frequencies may not be of interest (the cutoff frequency of waveguide, for example). In such cases, these resonant frequencies should be avoided when we design the source spectrum. One of the most commonly used excitation signals is the Gaussian pulse, due to its fast convergence in both time and frequency domains. However, a Gaussian signal contains a large direct component (DC), and is not needed in many simulations, thus its variations, Gaussian derivative and modulated Gaussian signals, are used more frequently. Of course, these two variations have nonzero derivatives at DC, thus the simulation result can still have DC spectrum, as will be discussed later.

In the following sections, we first discuss the time domain properties of excitation pulses, followed by the implementation of the local source. Sources on uniform transmission lines are special cases of local source, and their unique features are described in a separate section. We then discuss the power calculation, array excitation, and the implementation of plane wave sources.

4.2 TIME SIGNATURE

Sources in the FDTD algorithm can be described by their distributions in both the spatial and time domains. For the simulation to converge in a finite number of time steps, the source needs to be a band-limited source. Strictly speaking, a time-limited signal has infinite spectrum in the frequency domain, but as an engineering approach, we can treat the component of spectrum as zero if it is negligibly small. The Gaussian signal has an analytical expression in both the time and frequency domains, thus it is a very good candidate for the FDTD simulations. Since the iteration starts from a zero time step, we need to add a delay in the signal so that it is small enough at time zero and the truncation error can be neglected. The form of a Gaussian signal can be written as:

$$S_1(t) = Ae^{-\left(\frac{t-t_0}{\tau}\right)^2} \qquad (4.1)$$

where A is the signal amplitude, t_0 is the time delay, and τ is the damping factor. The frequency spectrum of the above signal is a Gaussian function and has a linear phase shift due to the time delay:

$$\tilde{S}_1(f) = \sqrt{\pi} A\tau e^{-(\pi \tau f)^2} e^{-j2\pi f t_0} \qquad (4.2)$$

The 3-dB frequency f_{3dB} of a Gaussian pulse is determined by the damping factor τ, namely, $f_{3dB} = 1/\pi\tau$. The drawback of the Gaussian signal is that it contains a large DC component. When used as a current source, it may result in charge accumulations for an open circuit, so that the field value will converge to a constant other than zero. Although such results are meaningful in physics, it may require extra effort in the data post-processing.

For this reason, two alternative Gaussian pulses are used more frequently in practice than the Gaussian signal itself. They are the Gaussian derivative signal and the modulated Gaussian signal. As the name indicates, Gaussian derivative signal is obtained by taking the derivative of (4.1):

$$S_2(t) = -2\frac{A}{\tau^2}(t-t_0)e^{-\left(\frac{t-t_0}{\tau}\right)^2} \qquad (4.3)$$

Its frequency spectrum can be written as:

$$\tilde{S}_2(f) = 2j\pi^{3/2} A\tau f e^{-(\pi\tau f)^2} e^{-j2\pi f t_0} \tag{4.4}$$

The Gaussian derivative signal is widely used because it does not have any DC component and covers a broad frequency band, while the spectrum is very simple. But this signal cannot be used in band-pass applications, thus it cannot avoid any specific frequencies that are expectably excluded from FDTD simulation. The modulated Gaussian signal, on the other hand, can overcome this problem. The signal is obtained by multiplying the Gaussian signal with a sinusoidal wave carrier:

$$S_3(t) = A e^{-\left(\frac{t-t_0}{\tau}\right)^2} \sin\left[2\pi f_0(t - t_0)\right] \tag{4.5}$$

where f_0 is the carrier frequency. Its spectrum is:

$$\tilde{S}_3(f) = \frac{\sqrt{\pi} A\tau}{2j} \left[e^{-[\pi\tau(f-f_0)]^2} - e^{-[\pi\tau(f+f_0)]^2} \right] e^{-j2\pi f t_0} \tag{4.6}$$

In this signal, the pulse width is controlled by the Gaussian signal, and the baseband signal is modulated to a center frequency, so we have an additional degree of freedom to shape our input pulse.

Other frequently used pulses in the FDTD simulation include the raised cosine signal and continuous sine wave [1]. The latter one is similar to the Gaussian, except that it is strictly time-limited, so that no truncation error is introduced. But its spectrum is more complicated than the Gaussian, which limits its application. As a matter of fact, for a linear system the form of the signal is not important, as long as it can cover the bandwidth of interest. This is because we are only interested in the frequency response, and the output waveform is just an intermediate step to calculate the system response. One of the major advantages of the FDTD method is that it can cover a broad frequency band in a single simulation; as long as we know the spectrum of the input pulse, we can obtain the system response of the entire spectrum. In very few situations the continuous sinusoidal wave might be used as the input signal to simulate the response at a single frequency. But since we have to start the signal from zero time step, the spectrum of the input signal itself is not limited to a signal frequency and is only an approximation. In addition, such simulations sacrifice the broadband feature of the FDTD method, and thus are rarely used in practice. We will not go into details of such excitation signals in this chapter.

4.3 LOCAL SOURCE

When an excitation is completely contained within the computational domain, it can be viewed as a local source. Local sources can be implemented using "hard" excitation or electric/magnetic current excitation. A hard source forces the field inside the source region to be a predefined time signal. Usually a local source has a spatial distribution of amplitude and a constant phase at all source points. Based on this assumption, the electric hard source can be expressed as:

$$E^n(i,j,k) = A(i,j,k)S(n\Delta t) \quad (4.7)$$

where $A(i, j, k)$ is the spatial distribution of amplitude, and $S(n\Delta t)$ is the value of the source signal at time step n. Apparently, the field value at the source points is only dependent on the spatial and temporal distribution of the source itself, and is not related to the field in the neighborhood. Since the voltage measured at both ends of such a source is independent of its load, it can be regarded an ideal voltage source in the circuit model. When the input pulse dies out, the field at the source points will be zero, and then it is possible to replace it with a PEC object. From the circuit point of view, this implies that the inner resistance of an ideal voltage source is equal to zero. In most of the simulations, however, such an artificial conductor should not exist, and the reflection from this conductor should not be present. This is not to say that the hard source is nonphysical, but it just means that such a source is not a good approximation to most problems. However, if we allow the fields in the source region to contribute to the iteration process even after the input pulse has died down to zero, as has sometimes been proposed in the literature, the result will not be meaningful even though we will no longer see artificial reflection. Similar to (4.7), we can also define a "hard" magnetic source by enforcing the magnetic field on a loop to be a predefined time signal, and this is equivalent to an ideal current source in the circuit model. The hard sources can be applied to situations where the source can be approximated as an ideal voltage or current source. Probe excitations in cavities or waveguides, for example, can be simulated by using this model.

Compared to the hard sources, both the electric and magnetic current sources are better realizations of the physical reality. When an electric current source is used for excitation, Maxwell's curl equation for the magnetic field can be written as:

$$\nabla \times \vec{H} = \varepsilon \frac{\partial \vec{E}}{\partial t} + \sigma \vec{E} + \vec{J}_i \quad (4.8)$$

where $\vec{J}_i$ is the source current density. The finite difference formulation of the above equation is given by:

$$\nabla \times \vec{H}^{n+1/2} = \varepsilon \frac{\vec{E}^{n+1} - \vec{E}^n}{\Delta t} + \sigma \frac{\vec{E}^{n+1} + \vec{E}^n}{2} + \vec{J}_{i,s}\left[\left(n+\frac{1}{2}\right)\Delta t\right] \quad (4.9)$$

which leads to the update equation for the electric field that reads:

$$\vec{E}^{n+1} = \frac{\varepsilon - 0.5\sigma\Delta t}{\varepsilon + 0.5\sigma\Delta t}\vec{E}^n + \frac{\Delta t}{\varepsilon + 0.5\sigma\Delta t}\nabla \times \vec{H}^{n+1/2} - \frac{\Delta t}{\varepsilon + 0.5\sigma\Delta t}\vec{J}_{i,s}\left[\left(n+\frac{1}{2}\right)\Delta t\right] \quad (4.10)$$

Note that the time instant of (4.10) is $(n+1/2)\Delta t$, and hence, the source current should also be sampled at this temporal sample point. The spatial sampling point of the source current is identical to that of the electric field, but the temporal sampling is one-half time step apart. We should also know that although the current is applied on a single grid line, it is actually a volume current density, and its basis function is the same as that of the electric field. The source current is then equal to the current density multiplied by the area of the cell. The magnetic current source has a similar expression and we omit the details of these sources.

In parallel processing, it is very common to have a local source straddle several subdomains, as shown in Figure 4.1.

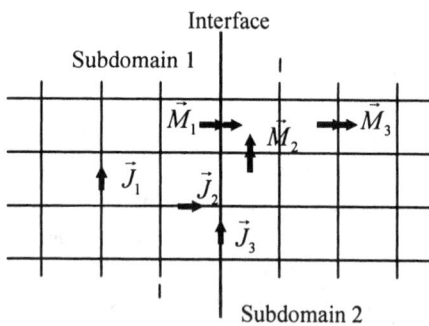

Figure 4.1 Electric/magnetic source in FDTD.

In Figure 4.1, the electric current J_3 and magnetic current M_1 are located on the domain boundary and should be applied to both subdomains. However, J_1 and M_3 only need to be included in their own domains. The magnetic current M_2 would be included in the update equation for subdomain-2, and the corresponding magnetic field would be passed to subdomain-1, hence, it should only be applied to subdomain-2. The electric current J_2 only affects the electric field in subdomain-1, and will not be passed to subdomain-2; consequently, it is sufficient to apply it to subdomain-1. An exception occurs in a situation where the result output requires the field averaging on the domain boundary. In this event, we include the point in subdomain-2 as well (i.e., apply the source in both subdomains).

Next, we demonstrate the implementation of the electric current source with a simple example. We set the cell size to be $\Delta x = \Delta y = \Delta z = \Delta = 0.25$ cm. The current source is applied to three grid lines in the z-direction, and the observation points are placed in a line moving away from the source, as shown in Figure 4.2.

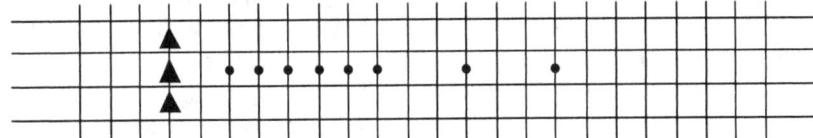

Figure 4.2 Electric current source in free space.

We first use the Gaussian pulse as the excitation. Since the Gaussian pulse has a DC component, it causes charge accumulation at both ends of the source. At the end of the iteration, the charge amount equals the time integration of the current:

$$Q = \int_{-\infty}^{t} I(t) dt = \Delta^2 \int_{-\infty}^{t} J(t) dt \qquad (4.11)$$

Due to the existence of the two charges, the field value at the observation points will converge to a constant, instead of to 0, at the end of the iteration. This is the static electric field due to the charges. Using charge obtained from (4.11), we calculate the static electric field using analytical formulations, and compare the field so derived with the numerical solutions at the observation points. The result is shown in Figure 4.3, and we can see that the agreement is very good in the region where the observation point is only a few cells away from the source.

The inset in Figure 4.4 illustrates the z-component of the electric field at the fifth observation point. To obtain the radiated field in the frequency domain, we need to carry out the Fourier transform of this signal. Since the signal has not been converged to 0, the fast Fourier transform (FFT) cannot be applied here. Instead, we divide the signal into two parts, a time-limited signal and a unit step signal:

$$\vec{E}(t) = \vec{E}_0(t) + \vec{E}_\infty u\left[t - (N + 1/2)\Delta t\right] \qquad (4.12a)$$

$$\vec{E}_0(t) = \begin{cases} \vec{E}(t), & t \le (N+1/2)\Delta t \\ 0, & t > (N+1/2)\Delta t \end{cases} \quad (4.12b)$$

where E_∞ is the electric field when the time is infinity, $u(t)$ is the unit step function, and N is the total time step. For the first signal, we can derive its spectrum using discrete Fourier transform (DFT), and add it to the analytic spectrum of the unit step signal. One-half time step increment in (4.12) is due to the sampling time instant, because the electric field is sampled at integral time steps, and its temporal basis function is extended by one-half time step in both directions. After carrying the above operation, we compare, in Figure 4.4, the numerical result with the analytical solution calculated from an electrical dipole with a length of 3-D and an amplitude of JD^2. We can see that the agreement between the FDTD and the analytic solution is very good. The amplitude J is frequency-dependent, and is determined by the excitation source, represented by (4.2) in this case.

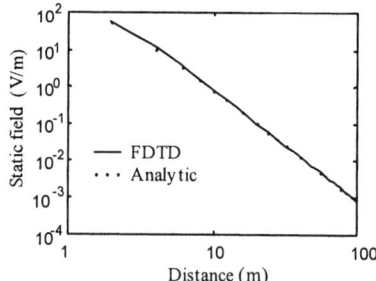

Figure 4.3 Static field obtained in FDTD.

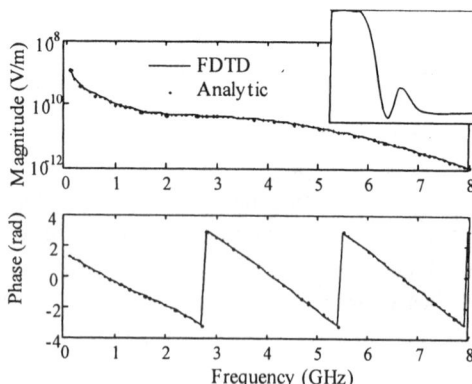

Figure 4.4 Radiated field due to a Gaussian source.

From this example, we can see that the FDTD can be applied to many DC problems. Of course, it is not as convenient and efficient as other circuit solvers, but for some extra broadband problems such as IC, package, and board designs, the DC characteristics are required in addition to the high frequency responses. We can use the FDTD to solve such a problem in a single simulation instead of using different tools for different frequency bands. This feature is not available in most of the frequency domain solvers. In most of the circuit problems, there is no charge accumulation at the end of the simulation because there exist loops for the currents. But for antenna problems such as the example above, charges could be accumulated. In such cases, we can use (4.12) and the DFT to calculate the frequency spectrum, but the efficiency is much lower than when we are able to use the FFT instead. Because of this, it is preferable to use excitation pulses that have no DC components.

In the second example, we use the modulated Gaussian signal as the excitation source, while the rest of the configurations remain the same as in the first example. At the same observation point, the numerical solution is compared with the analytical radiated field, as shown in Figure 4.5.

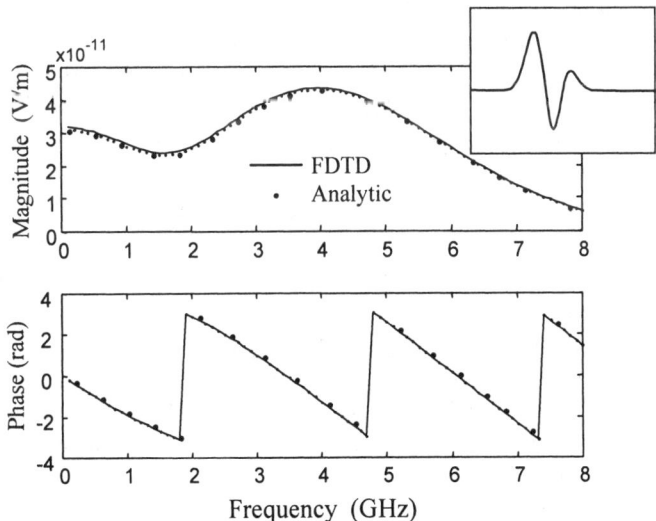

Figure 4.5 Radiated field when modulated Gaussian pulse is used.

We note that although the source current does not have a DC component, the radiated field does. This is because the charge is time integration of the current, and even though it did not accumulate at the end, the spectrum of the charge signal would still have a DC component, and this would result in static field distribution at the observation point. This field attenuates as the cube of the distance, so it can only be observed close to the source. An equivalent explanation is that the $1/f$ in the near-field radiation formula approaches to infinity at 0 frequency, and it cancels out with the infinitely small value in the spectrum of the source, resulting in a nonzero constant. The $1/f$ infinity also appears in the first example, inside the spectrum of the unit step function in (4.12).

For a resonant structure excited by a local source, the time domain signature usually exhibits a slow convergence except when the contents of the source spectrum are very small at the resonant frequencies of the system. An efficient approach to reducing the oscillation lingering on in the time domain is to introduce an internal resistance in the local source.

4.4 SOURCES FOR UNIFORM TRANSMISSION LINES

Uniform transmission lines are guided-wave structures that can support some kinds of modal fields in their cross sections. To derive the network parameters for a particular mode, the excitation source distribution should correspond exactly to that mode. Hence, we must employ local sources that have a modal character when we deal with uniform transmission lines. For canonical shapes such as rectangular or circular waveguides or coaxial cables, the distribution can be obtained analytically. But for an arbitrary cross section an analytical solution usually does not exist, and the modal solution must be derived numerically, hence, we will focus on this case. The modes in a guide with an arbitrary cross section can be categorized into two types, namely, the TEM and quasi-TEM waves supported by two or more conductors, and the TE and TM waves in closed guides. The latter also includes higher-order modes in parallel plate transmission lines. The field distribution in a TEM guide can be obtained by solving for the scalar potential in the cross section, while the modal distribution associated with the TE or TM waves can be obtained by solving for the eigenvalue problem for the propagation constants, and then deriving the fields that correspond to these wave numbers. Although Heaviside realized the possibility of the electromagnetic waves propagating in metal tubes as early as 1893, he assumed that the propagation requires at least two conductors. In 1897, Lord Rayleigh proved the possibility of propagation in rectangular and circular waveguides, and that the number of TE and TM waves with different cutoff frequencies can be infinite.

A TEM mode has no field components along the direction of propagation, and the voltage and current, as well as the characteristic impedance, can be uniquely defined for this mode. In such cases, the tangential field in the cross section of the guide can be expressed in terms of scalar potential functions ϕ and ϕ_H [2, 3]:

$$\vec{E} = \nabla \phi \tag{4.13a}$$

$$\vec{H} = \nabla \phi_H \tag{4.13b}$$

The potentials related to the electric and magnetic fields are not independent, thus we need only to solve for the electric potential. To better adapt the source excitation to a finite difference scheme, we sample the potential at the grid points of the FDTD mesh. This enables us to write (4.13) in a finite difference form:

$$E_x(i+1/2, j, k) = \frac{\phi(i, j, k) - \phi(i+1, j, k)}{\Delta x} \tag{4.14a}$$

$$E_y(i, j+1/2, k) = \frac{\phi(i, j, k) - \phi(i, j+1, k)}{\Delta y} \tag{4.14b}$$

Since the potential on a conductor is a constant, we can assign different potential on different conductors, and view them as known quantities. This is because the potential itself can have relative values without affecting the field distribution. In the space outside the conductors, there should not be any free charge, so we have:

$$\nabla \cdot \vec{D} = 0 \tag{4.15}$$

We write the above equation in a finite difference form, and substitute (4.14) in it to get the following equation for the potential ϕ:

$$\frac{\varepsilon_x(\phi(i, j, k) - \phi(i+1, j, k)) - \varepsilon_x(\phi(i-1, j, k) - \phi(i, j, k))}{\Delta x^2}$$
$$+ \frac{\varepsilon_y(\phi(i, j, k) - \phi(i, j+1, k)) - \varepsilon_y(\phi(i, j-1, k) - \phi(i, j, k))}{\Delta y^2} = 0 \tag{4.16}$$

If we move the known quantities, namely the assigned potential on the conductors to the right-hand side of (4.16), we will obtain a set of linear equations for the dicretized values of the potential distribution to be solved. The value Δx and Δy in (4.14) and (4.16) should be adjusted to the sampling points when nonuniform mesh is used. Likewise, the equations in (4.16) also need to be modified when a conformal technique is used though the solution process will remain the same.

When the number of conductors in the guide is greater than two, multiple TEM modes can be supported by the system of conductors (n modes for n conductors) [3]. In this case, we need to solve (4.16) more than once, assigning different potentials to each conductor each time. In general we do not obtain the modal distributions directly when we follow the above procedure, but only a linear combination of the modes, which we can separate by performing a singular value decomposition (SVD) for the matrix of the solution vectors. The SVD yields a set of orthogonal modes, but the modes we derive in this manner may not be the conventional modes used in classical analysis (for example, the even and odd modes in parallel microstrip lines), unless we further manipulate the solutions. In most of the practical applications, we can obtain the desired mode distributions by forcing the orientation of one field component for a specific mode (e.g., along one of the axes in the Cartesian system), which is consistent with the specification in the conventional FDTD algorithm.

When dealing with open transmission line structures, it becomes necessary to apply appropriate boundary conditions. Most common boundary conditions are the PEC, PMC, and absorbing boundary conditions, as we discussed earlier in Chapter 2. For problems with symmetries, using the PEC or PMC can help us reduce the number of unknowns as well as the simulation time. When the electric field is perpendicular to the plane of symmetry, a PEC boundary can be applied. The PEC boundary is a conductor and we can assign a known potential to it in a manner similar to other PEC objects. When the electric field is parallel to the plane of symmetry, a PMC boundary can be applied. The potential on the PMC boundary is unknown, hence, we must find a different way to handle this situation as explained below.

When solving (4.16) with a PMC boundary, we need to know the field (potential) values just outside the above boundary. Fortunately, we can obtain them through the use of imaging, noting that the electric field images by a PMC with the same polarity.

For the PML absorbing boundary condition, the simplest way to implement it is to write the PML equations as Maxwell's equations in complex coordinates [4]. This does not alter (4.16), except that the cell sizes (Δx and Δy) in the denominator now become complex numbers. The solution to this equation is also complex. However, for the purpose of the excitation we can use its real part in the FDTD.

In the following example, we solve for the mode in a pair of microstrip lines as shown in Figure 4.6. Since this is an open structure, the enclosing boundaries are set to be PML except for the PEC ground plane. We solve for equation (4.16) using complex coordinates, and the field distribution of the two resulting modes are plotted in Figure 4.7. The black line in the figures indicates the interfaces where the PML boundary condition is enforced.

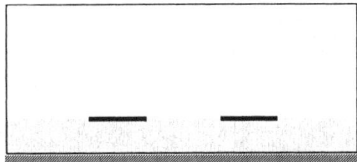

Figure 4.6 Parallel microstrip lines.

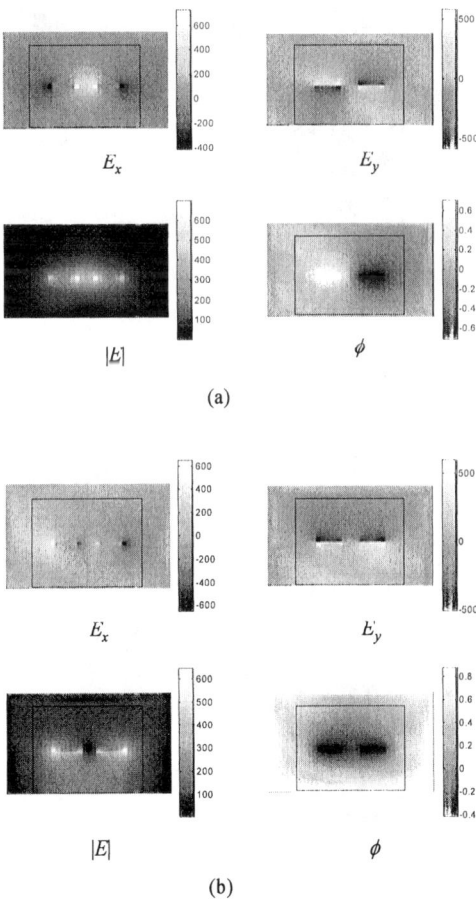

Figure 4.7 TEM modes for the parallel microstrip lines. (a) Odd mode. (b) Even mode.

Next, we insert a PEC boundary at the middle of the two microstrip lines (i.e., the plane of symmetry). The resulting field distribution is shown in Figure 4.8. We can see the resemblance between the solution derived with the PEC boundary and the odd mode obtained previously (see Figure 4.7). Next, we change the boundary to be PMC, and the result is the even mode, as shown in Figure 4.9.

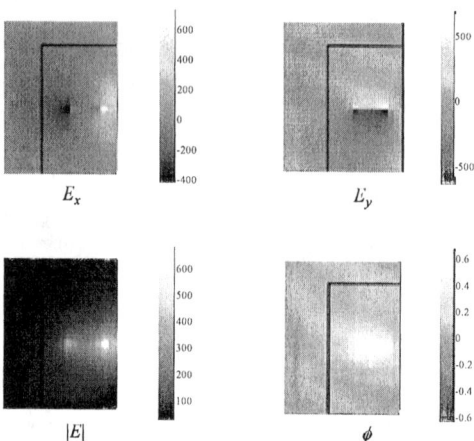

Figure 4.8 Odd mode when PEC boundary is applied.

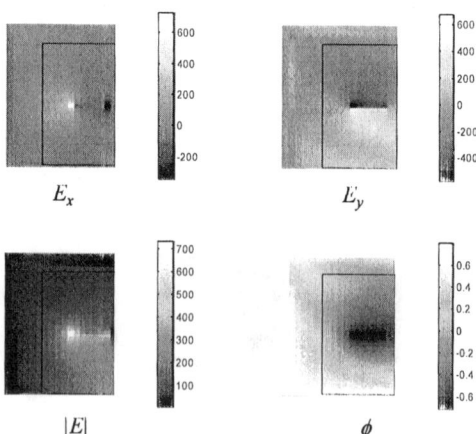

Figure 4.9 Even mode when PMC boundary is applied.

In addition to the TEM modes on parallel conductors, the other most commonly guided waves are the TE or TM modes in waveguide structures, including the higher-order modes on parallel conductors. For these modes, the field component in the longitudinal direction (i.e., transverse to the cross section is nonzero), thus (4.13) is no longer applicable. Instead, we write the finite difference equations for the frequency domain for the fields on the cross section, and solve for the cutoff frequencies and field distribution by extracting the eigenvalues and eigenvectors. Using the FDTD grid, we can write the frequency domain Maxwell's equation in a finite difference form:

$$\nabla \times \tilde{\vec{H}} = j\omega\varepsilon\tilde{\vec{E}} \tag{4.17}$$

Suppose a TE wave propagates along the z-direction, at the cutoff frequency of the mode, all the field components will be uniform in the z-direction, and only the horizontal variations of the field will contribute to the curl. Also the z-component of the electric field would be zero, and we can write the equation for $\tilde{E}_x$, $\tilde{E}_y$, and $\tilde{H}_z$ in a matrix form as:

$$\left(\nabla_H \times\right) \begin{bmatrix} \tilde{\vec{H}}_1 \\ \vdots \\ \tilde{\vec{H}}_m \end{bmatrix} = j\omega \begin{bmatrix} \varepsilon_1 & & 0 \\ & \ddots & \\ 0 & & \varepsilon_n \end{bmatrix} \begin{bmatrix} \tilde{\vec{E}}_1 \\ \vdots \\ \tilde{\vec{E}}_n \end{bmatrix} \tag{4.18a}$$

$$\left(\nabla_E \times\right) \begin{bmatrix} \tilde{\vec{E}}_1 \\ \vdots \\ \tilde{\vec{E}}_n \end{bmatrix} = -j\omega \begin{bmatrix} \mu_1 & & 0 \\ & \ddots & \\ 0 & & \mu_m \end{bmatrix} \begin{bmatrix} \tilde{\vec{H}}_1 \\ \vdots \\ \tilde{\vec{H}}_m \end{bmatrix} \tag{4.18b}$$

where n and m represent the number of unknown tangential electric fields and perpendicular magnetic fields, respectively. The matrix $\left(\nabla_H \times\right)$ is an $n \times m$ matrix, and $\left(\nabla_E \times\right)$ is an $m \times n$ matrix, derived by using the difference version of the curl. If we take the material property matrix to the left and substitute the first equation into the second one, it would lead to the wave equation in the finite difference scheme that reads:

$$\begin{bmatrix} \varepsilon_1 & & 0 \\ & \ddots & \\ 0 & & \varepsilon_n \end{bmatrix}^{-1} (\nabla_H \times) \begin{bmatrix} \mu_1 & & 0 \\ & \ddots & \\ 0 & & \mu_m \end{bmatrix}^{-1} (\nabla_E \times) \begin{bmatrix} \tilde{\vec{E}}_1 \\ \vdots \\ \tilde{\vec{E}}_n \end{bmatrix} = \omega^2 \begin{bmatrix} \tilde{\vec{E}}_1 \\ \vdots \\ \tilde{\vec{E}}_n \end{bmatrix} \quad (4.19)$$

or in a compact form:

$$\mathbf{A}\tilde{\vec{E}} = \omega^2 \tilde{\vec{E}} \quad (4.20)$$

Since the above equation is homogenous, its solution would be the eigenmode of the system (i.e., a distribution that resonates in the structure). This is exactly what happens to a TE mode at its cutoff frequency. Examining (4.20) we can see that it is an eigenvalue problem, where the eigenvalues of **A** are ω^2, and $\omega/2\pi$ is the cutoff frequency. The corresponding eigenvector provides the information on the field distribution for the corresponding TE mode. In practice, it is preferable to solve for the perpendicular magnetic field instead of the electric field, because the number of unknowns we have to deal with it is approximately half. We should point out that we may get a few eigenvalues whose amplitudes are very small, but they would be numerical artifacts and should be discarded as noise. Even if some of the cutoff frequencies are above the highest frequency of the source spectrum, we should nonetheless retain a few of them in many cases, because they may still be important in many coupling problems. The procedure for solving the TM mode is exactly the same as that for the TE mode, except that the tangential magnetic field and the perpendicular electric field now become the unknown. For either the PEC/PMC boundary or the PML absorbing boundaries, the method of handling is the same as that for the TEM mode extraction. If there are degenerated modes in the system, that is, if several modes have the same cutoff frequency, they can be separated using the same method as that used for the TEM modes.

In the following example, we consider a circular waveguide with a 0.2-inch radius. The finite difference cell size is taken to be 0.01 inch. By solving for the eigenvalue and eigenvector of the matrix **A** in (4.20), we can derive a series of TE modes, and we illustrate the first of these in Figure 4.10. This is a degenerate mode, and the two modes are separated by forcing the x-component of the electric fields to be orthogonal. The resulting field distribution agrees with the analytical TE_{11} mode in the circular waveguide, and the separation of the degenerated modes also agrees with the expected result. The cutoff frequency obtained from the numerical method is 16.8 GHz, while that of the actual TE_{11} mode is 17.3 GHz. This error is due to the staircasing approximation in the FDTD, and could be reduced by applying the conformal technique.

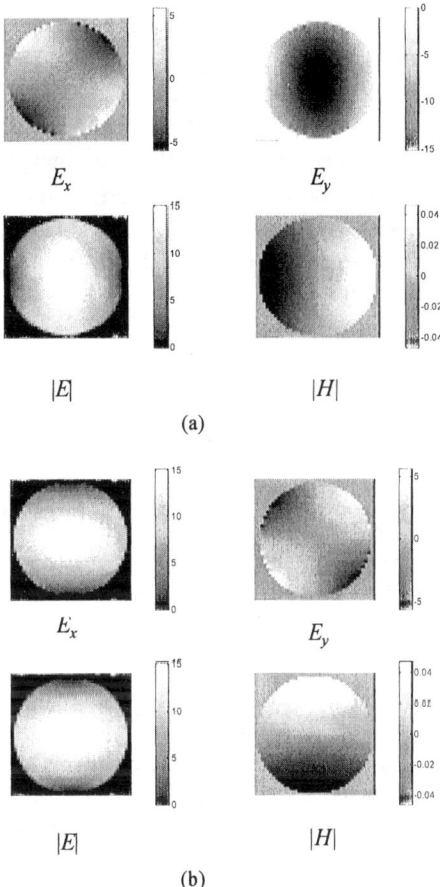

Figure 4.10 Numerical solution of the first mode in a circular waveguide. (a) Degenerated mode one. (b) Degenerated mode two.

Next, we introduce the methodology for mode extraction on uniform transmission lines that have arbitrary cross sections. If we use the method that we will outline below, our simulation will not be limited to some special cases, and we would not need to design a source distribution for individual transmission lines. Hence, the source implementation can be fully automated. From the above example we can see that, in some cases, the FDTD solution can be different from the analytical solution for canonical problems due to the approximation of the geometry. In such cases, applying the analytical field distribution may not be a good choice, because the solution would be inconsistent with the discretized system, and is liable to affect the accuracy of the results. For instance, if we use the analytical TE_{11} mode distribution to excite the waveguide in the FDTD in the above example, higher-order modes could be excited. Therefore, for every port on the transmission line, it is suggested that the mode distribution be derived numerically, and the corresponding numerical cutoff frequency be used to compute the wave impedance and propagation constant. To excite a certain mode, we can apply its distribution at a cross section of the guide, and two traveling waves that have the same amplitude but propagate opposite directions. The electric or magnetic currents associated with the source can be obtained directly from the field distribution:

$$\vec{J}_s = \hat{n} \times \vec{H} \qquad (4.21a)$$

$$\vec{M}_s = \vec{E} \times \hat{n} \qquad (4.21b)$$

where $\hat{n}$ is the unit vector along the direction of propagation and $\vec{J}_s$ and $\vec{M}_s$ are the equivalent surface electric and magnetic current densities, respectively. Either the electric or magnetic current can be used independently to excite the mode. Since the source current in the FDTD is a volume current, the equivalent surface current is approximated by a volume current in one cell:

$$J_s = J\Delta \qquad (4.22)$$

The procedure for dealing with the magnetic current is also similar.

We should note that for the TE and TM waves, the wave impedances of the transmission lines are frequency-dependent. The impedance of the TE wave, for example, approaches infinity at the cutoff frequency. If we use the electric current source to excite a TE mode, the impedance term will appear as a coefficient of the electric field [5–9], resulting in very large values of the electric field in the frequency domain, or slow convergence in the time domain. Using a magnetic current source instead will not cause such a problem, because the impedance appears in the denominator. Thus, as a general guideline, the TE modes should be excited by using the magnetic current sources, and the TM modes with the electric current sources. For a TE wave propagating in the z-direction, we can see from (4.21) that the x-component of the magnetic current is equal to the y-component of the electric field. In the FDTD, M_x and E_y are sampled at the same point in the x-y plane, while being separated by a one-half cell in the z-direction. When the electric field distribution is known, the implementation of magnetic current is relatively straightforward.

At the output port, the frequency domain field distribution is a linear combination of all the modes:

$$\tilde{\vec{E}} = \sum_n A_n \tilde{\vec{E}}_n \qquad (4.23)$$

where $\tilde{\vec{E}}_n$ is the field distribution of the nth mode, and A_n is its amplitude. Since all the mode distributions are orthogonal, the modal amplitude can be easily obtained by projecting the field distribution to the corresponding modal distribution (i.e., by using):

$$A_n = \frac{\langle \tilde{\vec{E}}, \tilde{\vec{E}}_n \rangle}{\langle \tilde{\vec{E}}_n, \tilde{\vec{E}}_n \rangle} \qquad (4.24)$$

Note that when a port is located very close to a discontinuity, a higher-order mode may also contaminate the field distribution, and this should be kept in mind while extracting the network parameters.

4.5 POWER INTRODUCED BY LOCAL EXCITATION SOURCES

The power introduced into the system by an excitation source is not only dependent on the source itself, but also on the entire system that it excites. Thus the power calculation requires the output of the FDTD simulation. We can compute either the instantaneous power or the average power at the source point, but by and large, the instantaneous power is not of interest to us, and the time domain instantaneous power does not have a direct Fourier relationship with the power spectrum in the frequency domain. Thus we only discuss the calculation of the average power at a certain frequency. Depending on the physical quantities of the source, the corresponding output quantities required in power calculation are also different. Hence, both the source and the output need to be transformed into the frequency domain to compute the power.

The circuit model for a hard source is an ideal voltage or current source. The output power is equal to the real part of the product of voltage and the conjugate of current:

$$P_{in} = \frac{1}{2} \mathrm{Re}\{\tilde{U}\tilde{I}^*\} \qquad (4.25)$$

For a hard electric source, the electric field at the source point is known, and the voltage is obtained by a line integration of the field. The current can be measured by a loop integration of the magnetic field on the cross section of the source. For a hard magnetic source, the current is known and the voltage is calculated by the line integral of the electric field.

For an electric current source, the frequency domain power is expressed as [6]:

$$P_{in} = \frac{1}{2} \iiint \mathrm{Re}\left\{ \tilde{\vec{J}}_{in} \cdot \tilde{\vec{E}}^* \right\} dv \qquad (4.26)$$

where $\tilde{\vec{J}}_{in}$ is the source current density whose spectrum is known. $\tilde{\vec{E}}$ is the electric field observed at the source point, and its spectrum is obtained via Fourier transform. Note that the field in (4.26) is not only related to the source, but also to the entire system, and can only be obtained through the FDTD simulation. The volume integration of the above equation involves the use of the expansion of the current or the field basis functions. Even when the current is applied on a single grid line, the integration will still involve one-half cell if all the four directions are perpendicular to the source. The power obtained form (4.26) is the total power at the source point, and the internal impedance of the source itself will consume part of the power, which is given by:

$$P_s = \frac{1}{2} \iiint \sigma |E|^2 \, dv \qquad (4.27)$$

The output power of the source is then the difference between the total power and the power consumed at the source:

$$P_{out} = P_{in} - P_s \qquad (4.28)$$

The ideal voltage or current sources do not consume any power themselves. Thus, the power in (4.25) is the output power of the hard source. The total power of a magnetic current source is:

$$P_{in} = \frac{1}{2} \iiint \mathrm{Re}\left\{ \tilde{\vec{M}}_{in} \cdot \tilde{\vec{H}}^* \right\} dv \qquad (4.29)$$

And the self-consumed power is:

$$P_s = \frac{1}{2} \iiint \sigma_M |H|^2 \, dv \qquad (4.30)$$

The output power is also the difference of the two. When the source is placed on the cross section of a uniform transmission line, two traveling waves will be excited, each carrying half of the output power, and we should be cognizant of this fact. Actually, when the modal distribution and the propagation constant are known, we can calculate the power from the field amplitudes at two points on the transmission line.

4.6 PHASE DIFFERENCE AND TIME DELAY

When multiple sources exist in the computational domain, they could in general have phase differences and, in the FDTD, this phase difference is implemented through a time delay. Suppose the *n*th source in the frequency domain is given by:

$$\tilde{J}_n = |\tilde{J}_n| e^{j\phi_n} \tag{4.31}$$

From the properties of the Fourier transforms we have

$$\Re\{S(t-t_n)\} = \Re\{S(t)\} e^{-j2\pi f t_n} \tag{4.32}$$

We know that when the phase difference between two sources is $\Delta\phi = \phi_2 - \phi_1$ for a given frequency f_0, the corresponding time delay is:

$$\Delta t = t_2 - t_1 = -\frac{\Delta\phi}{2\pi f_0} \tag{4.33}$$

The above equation is only valid at one frequency, so for a given time delay, the phase difference varies with frequencies. However, when the phase difference is linearly dependent on the frequency — as is often the case — we can use a fixed time delay for all frequencies. Another thing to keep in mind is that the reference pulse measuring the time delay always has a zero value at the beginning of the FDTD simulation, hence, any meaningful time delay of the excitation should be greater than zero. Note that the time delay is a relative quantity in (4.33), and its absolute value is constrained by the condition mentioned above.

Multiple source excitations are typically utilized in array problems. For a multiple port network, we should add the source at one port at a time, and perform a simulation for each source, to derive the network parameters. However, for large array problems, such a scheme is usually computationally very intensive, and we are sometimes more interested in the far-field distribution. In such cases, all the sources can be turned on concurrently, and the phase differences can be controlled by the time delay. For an ideal phased array, the phase of the sources can be obtained through the Fourier relationships between the spatial and spectral domains. Suppose the scan angle is (θ, φ), the corresponding components of the wave number are $k_x = k_0 \sin\theta\cos\varphi$, $k_y = k_0 \sin\theta\sin\varphi$, and $k_z = k_0 \cos\theta$, the phase for each excitation source is then:

$$\phi_n = -k_x x_n - k_y y_n - k_z z_n \tag{4.34}$$

Substituting (4.34) into (4.33), we obtain the time delay to be:

$$\Delta t = \frac{(x_n \sin\theta\cos\varphi + y_n \sin\theta\sin\varphi + z_n \cos\theta)}{c} \tag{4.35}$$

From (4.35), we can see that the time delay is only dependent on the scan angle. Once this delay is fixed, the scan angle is the same for all frequencies. This is easier to understand from the spatial domain point of view, because the time delay in (4.35) actually corresponds to the travel time from the elements to an equal-phase plane.

4.7 PLANE WAVE SOURCES

Up to this point, our discussions on source implementation are all based on the total field formulation, in which the unknowns are the total electric and magnetic fields. In a scattering problem, the incident wave is a planewave incident from infinity, and cannot be implemented in a finite space using the total field formulation. But since the incident field is already known, we can choose the scattered field as our unknown. Note that the scattered field emanates from the object, at the time instant when the incident wave impinges upon the object.

There are two types of implementations used to handle the scattering problems. The first one of these is the scattered field formulation [7], in which the unknowns are the scattered fields in the entire computational domain. The second choice is the total/scattered field formulation [1, 8, 9], which the unknowns are the total fields in part of the computational domain, and the scattered field in the rest. Both of the formulations have their unique features and advantages (as well as limitations), but the total/scattered field is easier to implement and is thus more widely used.

Let us start with a discussion of the scattered formulation first. The scattered field is produced by the induced electric and magnetic currents on the object, and the incident field is the radiated field produced by the source in free space. The sum of the two is the total field [10]:

$$\vec{E}^t = \vec{E}^i + \vec{E}^s \quad (4.36a)$$

$$\vec{H}^t = \vec{H}^i + \vec{H}^s \quad (4.36b)$$

where $\vec{E}^t$ is the total field, $\vec{E}^i$ is the incident field, and $\vec{E}^s$ is the scattered field. The total field satisfies Maxwell's equations:

$$\nabla \times \vec{H}^t = \varepsilon \frac{\partial \vec{E}^t}{\partial t} + \sigma \vec{E}^t \quad (4.37a)$$

$$\nabla \times \vec{E}^t = -\mu \frac{\partial \vec{H}^t}{\partial t} - \sigma_M \vec{H}^t \quad (4.37b)$$

Since the incident field propagates in free space, it satisfies Maxwell's equations in free space:

$$\nabla \times \vec{H}^i = \varepsilon_0 \frac{\partial \vec{E}^i}{\partial t} \quad (4.38a)$$

$$\nabla \times \vec{E}^i = -\mu_0 \frac{\partial \vec{H}^i}{\partial t} \quad (4.38b)$$

Substituting (4.36) and (4.38) into (4.37), we can obtain the following equations for the scattered field:

$$\nabla \times \vec{H}^s = \varepsilon \frac{\partial \vec{E}^s}{\partial t} + \sigma \vec{E}^s + (\varepsilon - \varepsilon_0) \frac{\partial \vec{E}^i}{\partial t} + \sigma \vec{E}^i \quad (4.39a)$$

$$\nabla \times \vec{E}^s = -\mu \frac{\partial \vec{H}^s}{\partial t} - \sigma_M \vec{H}^s - (\mu - \mu_0) \frac{\partial \vec{H}^i}{\partial t} - \sigma_M \vec{H}^i \quad (4.39b)$$

Writing (4.39) in the finite difference formulation, we can derive the update equations in the total field formulation:

$$\vec{E}^{s,n+1} = \frac{\varepsilon - 0.5\Delta t\sigma}{\varepsilon + 0.5\Delta t\sigma} \vec{E}^{s,n} + \frac{1}{\varepsilon + 0.5\Delta t\sigma} \nabla \times \vec{H}^{s,n+1/2}$$
$$- \frac{\Delta t}{\varepsilon + 0.5\Delta t\sigma} \left[(\varepsilon - \varepsilon_0) \frac{\partial \vec{E}^i}{\partial t} \bigg|^{n+1/2} + \sigma \vec{E}^{i,n+1/2} \right] \quad (4.40a)$$

$$\vec{H}^{s,n+1/2} = \frac{\mu - 0.5\sigma_M \Delta t}{\mu + 0.5\sigma_M \Delta t} \vec{H}^{s,n-1/2} - \frac{1}{\mu + 0.5\sigma_M \Delta t} \nabla \times \vec{E}^{s,n}$$

$$- \frac{\Delta t}{\mu + 0.5\sigma_M \Delta t} \left[(\mu - \mu_0) \frac{\partial \vec{H}^i}{\partial t} \bigg|^n + \sigma_M \vec{H}^{i,n} \right] \quad (4.40\text{b})$$

We note that (4.40a) and (4.40b) have the same update equations as the conventional FDTD, except for the two terms related to the incident field. Whenever the material is different from that of the free space, these two terms need to be added to the update equations. Inside the perfect conductors, the total field is equal to zero, and from (4.36) we have:

$$\vec{E}^{s,n} = -\vec{E}^{i,n} \quad (4.41)$$

Note that the update equation given in (4.40) is derived at the time instant $(n+1/2)\Delta t$; hence, the incident field and its derivative should be sampled at this time instant. On the other hand, (4.41) is based on the time step $n\Delta t$, and the incident field should be taken at integral time steps. It would be useful to clarify the physical meanings of these two equations. For an even symmetric system truncated by a PMC boundary, the scattered field on the PMC boundary equals the incident field, namely, $\vec{E}^{s,n} = \vec{E}^{i,n}$.

The starting time of the scattered field formulation should be the time instant when the incident pulse enters the global computational domain. Depending on the incident direction, the time delay of the excitation pulse should be adjusted in each subdomain, so that the incident field will always start from the nearest corner through which it enters. Suppose a plane wave has an incident angle of (θ, φ), and the incident pulse is $S(t)$. Then the incident field at any point in the computational domain can be expressed as:

$$\vec{E}(x,y,z) = \left(\hat{\theta} E_\theta + \hat{\varphi} E_\varphi\right) S\left(t + \frac{x\sin\theta\cos\varphi + y\sin\theta\sin\varphi + z\cos\theta}{c}\right) \quad (4.42)$$

To force the incident field in the entire domain to be zero at the zero time step, we need to add a time delay so that the incident pulse is expressed as $S(t-t_d)$, where t_d satisfies

$$t_d \geq \frac{x\sin\theta\cos\varphi + y\sin\theta\sin\varphi + z\cos\theta}{c} \quad (4.43)$$

in the entire domain. In practice we can take the equal sign in the above equation, because the scattered field is zero in the entire domain before the start of the incident pulse, and the delay only extends the zero signal. For different incident angles, the expression given in (4.43) could be different, and hence, the time delay can also be different. In the 2-D example shown in Figure 4.11, for example, the incident wave 1 corresponds to $\sin\theta = 1$, and both $\cos\varphi$ and $\sin\varphi$ have positive values. However, for the incident wave 2, $\cos\varphi$ is positive and $\sin\varphi$ is negative, so that the time delay becomes:

$$t_d = \frac{x_{\max}\cos\varphi + y_{\min}\sin\varphi}{c} \qquad (4.44)$$

where $(x_{\max}, y_{\min})$ is the reference point.

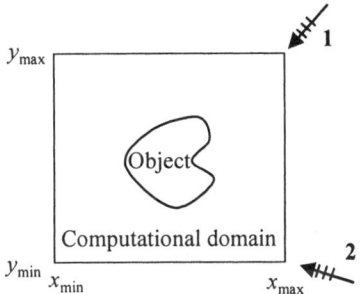

Figure 4.11 Incident wave in the scattered field formulation.

In the scattered field formulation, the incident source is applied in the entire computational domain. The advantage is that the source values are obtained analytically, and are not affected by the numerical dispersion of the finite difference cells, and this can help reduce the error in the FDTD simulation. The disadvantage is that the incident field and its derivative need to be calculated at every field point, which is more time consuming. An alternative solution is the total/scattered field formulation, which is depicted in Figure 4.12. In such a scheme, the computational domain is divided into the total and scattered field zones. The scatterers are included entirely within the total field zone, and the fields in this zone satisfy (4.37). On the other hand, the fields in the scattered field zone are governed by (4.39). However, since the scattered field zone is in free space, the two terms related to the incident field are zero in the update equations, so this equation is the same as that in the conventional FDTD. The total and scattered fields can be stored in the same data structure, as long as we can identify them.

On the boundaries of the two zones, the incident field is used to compensate for the field from the other zone. When the zone boundary coincides with the FDTD grid plane, the tangential electric field and perpendicular magnetic field are the total fields, and the tangential magnetic field and perpendicular electric field one-half cell apart from the boundary are the scattered fields. In Figure 4.12, the update equation for the y-component of the electric field is [11]:

$$E_y^{t,n+1}(i,j+1/2,k) = \frac{\varepsilon - 0.5\Delta t\sigma}{\varepsilon + 0.5\Delta t\sigma} E_y^{t,n}(i,j+1/2,k)$$

$$+ \frac{\Delta t}{\varepsilon + 0.5\Delta t\sigma} \left[\begin{array}{c} \dfrac{H_x^{t,n+1/2}(i,j+1/2,k+1/2) - H_x^{t,n+1/2}(i,j+1/2,k-1/2)}{\Delta z} \\ - \dfrac{H_z^{s,n+1/2}(i+1/2,j+1/2,k) - H_z^{t,n+1/2}(i-1/2,j+1/2,k)}{\Delta x} \end{array} \right] \quad (4.45)$$

$$- \frac{\Delta t}{\varepsilon + 0.5\Delta t\sigma} \frac{H_z^{i,n+1/2}(i+1/2,j+1/2,k)}{\Delta x}$$

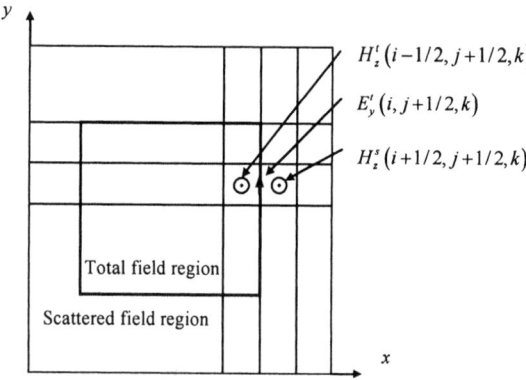

Figure 4.12 Illustration of the total/scattered field formulation.

The magnetic field at one-half cell outside the interface between the total and scattered field region can be updated:

$$H_z^{s,n+1/2}(i+1/2,j+1/2,k) = \frac{\mu - 0.5\sigma_M \Delta t}{\mu + 0.5\sigma_M \Delta t} H_z^{s,n-1/2}(i+1/2,j+1/2,k)$$

$$- \frac{\Delta t}{\mu + 0.5\sigma_M \Delta t} \left[\begin{array}{c} \dfrac{E_y^{s,n}(i+1,j+1/2,k) - E_y^{t,n}(i,j+1/2,k)}{\Delta x} \\ - \dfrac{E_x^{s,n}(i+1/2,j+1,k) - E_x^{s,n}(i+1/2,j,k)}{\Delta y} \end{array} \right] \quad (4.46)$$

$$- \frac{\Delta t}{\mu + 0.5\sigma_M \Delta t} \frac{E_y^{i,n}(i,j+1/2,k)}{\Delta x}$$

The update equations on the other planes are similar to the above, and will not be elaborated on here, though it should be fairly straightforward to derive them.

The start time in the total/scattered field formulation is treated in the same way as in the scattered field formulation, except that since we know the scattered field zone is in free space, the delay condition (4.43) only needs to be satisfied in the total field zone. In the total/scattered field formulation, the incident field only appears on the interface between the two zones, thus the simulation is less accurate than the scattered field formulation. But the advantages of this formulation are obvious because we neither need to insert the source fields in the entire computational domain nor calculate any derivatives, and both of these serve to reduce the simulation time. In receiving antenna problems, we are more interested in the total field at the output port, although its relationship with the scattered field is relatively simple. In any event, it is still easier to use the total/scattered field formulation.

We should note that the two formulations mentioned above are not limited to plane wave sources. They can also be applied to arbitrary sources, including local sources, because the incident field due to a local source can also be calculated analytically. In fact, when the local source is not placed on the FDTD grid, the scattered field or total/scattered field formulations yield better solutions to the problem than the direct local source simulation.

REFERENCES

[1]   A. Taflove and S. Hagness, *Computational Electromagnetics: The Finite-Difference Time-Domain Method*, 2nd ed., Artech House, Norwood, MA, 2000.

[2]   R. Collin, *Field Theory of Guided Wave*, Wiley-IEEE Press, New York, 1990.

[3]   D. Pozar, *Microwave Engineering*, 2nd ed., John Wiley & Sons, New York, 1998.

[4]   W. Chew, J. Jin, and E. Michielssen, "Complex Coordinate Stretching as a Generalized Absorbing Boundary Condition," *Microwave and Optical Technology Letters*, Vol. 15, No. 6, August 1997, pp. 363-369.

[5]   R. Harrington, *Field Computation by Moment Methods* (Macmillan Series in Electrical Science), Macmillan, New York, 1968.

[6]   R. Harrington, *Time-Harmonic Electromagnetic Fields*, Wiley-IEEE Press, New York, 2001.

[7]   T. Huang, B. Houshmand, and T. Itoh, "Efficient Modes Extraction and Numerically Exact Matched Sources for a Homogeneous Waveguide Cross-Section in a FDTD Simulation," *IEEE MTT-S Int. Microwave Symp.*, San Diego, CA, Vol. 1, May 1994, pp. 31-34.

[8] S. Wang and F. Teixeira, "An Equivalent Electric Field Source for Wideband FDTD Simulations of Waveguide Discontinuities," *IEEE Microwave and Wireless Components Letters*, Vol. 13, No. 1, January 2003, pp. 27-29.

[9] J. Schneider, C. L. Wagner, and O. Ramahi, "Implementation of Transparent Sources in FDTD Simulations," *IEEE Transactions on Antennas and Propagation*, Vol. 46, August 1998, pp. 1159-1168.

[10] K. Kunz and R. Luebbers, *The Finite Difference Time Domain Method for Electromagnetics*, CRC Press, Boca Raton, FL, 1993.

[11] K. Mashankar and A. Taflove, "A Novel Method to Analyze Electromagnetic Scattering of Complex Objects," *IEEE Transactions on Electromagnetic Compatibility*, Vol. 24, November 1982, pp. 397-405.

Chapter 5

Data Collection and Post-Processing

The direct outputs in the FDTD simulation are the time domain electric and/or magnetic fields at the observation points located in the computational domain, or the voltages and currents obtained by integrating these fields. In practice, we are also interested in several other quantities that are not as easy to obtain. In this chapter, we discuss some post-processing techniques related to the near fields, circuit parameters, and far-field parameters.

5.1 TIME-FREQUENCY TRANSFORMATION

As mentioned above, the unknowns in the FDTD simulation are the time domain electric and magnetic fields in the computational domain, and all other physical quantities are derived from these two [1–3]. For example, the voltage can be calculated by integrating the electric field along a voltage path, and the current can be obtained using a loop integral of the magnetic field. Most of the time we are interested in the frequency response of the system, and this is obtained by Fourier transforming the time domain signal. For scattering or radiation problems, the far fields are obtained from the spatial Fourier transform of the equivalent current that are derived from the near fields. To ensure that the results from these transforms are accurate, we need to have a more detailed understanding of the sampling in the FDTD algorithm.

In the FDTD, the temporal and spatial sampling points of the electric and magnetic fields are staggered. The fields are second-order accurate at their sampling points when we use central differencing and a uniform mesh. The field values are not defined outside the sampling points, and are usually approximated using linear interpolation. Such interpolations are no longer second-order accurate, and thus should be avoided unless really necessary. A typical example of a situation where we need to carry out such interpolation is when we want to measure the electric and magnetic fields at the same point. In Figure 5.1, for example, we need to measure the voltage between the microstrip and the ground plane, and the current on the microstrip. The voltage is a line integration of the electric field along the *y*-direction, and the integration line overlaps with the grid line. The current, however, is a loop integration of the magnetic field, and the integration path is one-half cell away in the *z*-direction from the electric field sampling point. At high frequencies, one-half cell difference can cause noticeable errors in the result. To reduce this effect, we need to measure the current at both sides of the electric field sampling point, and take the average to be the current at that sampling point, which is equivalent to a linear interpolation of the magnetic field at the sampling point of the electric field. Although the voltage and current are now measured at the same location, their temporal sampling point is still one-half time step apart. This difference can also be compensated by interpolation, but only if we wish to derive the frequency response. It is simpler as well as more effective to correct for this difference during the process of the Fourier transform, rather than by using interpolation.

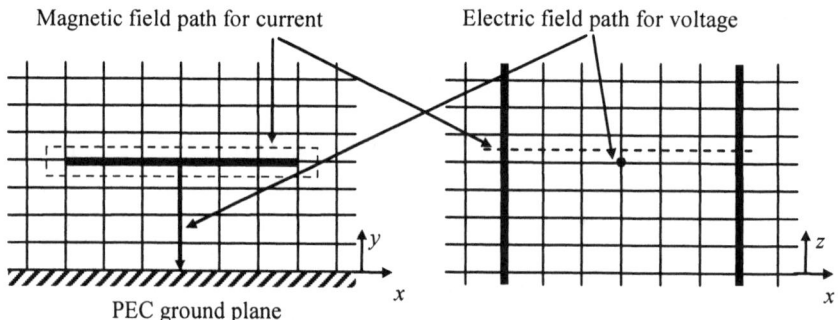

Figure 5.1 Voltage and current measurement on a microstrip line.

Fourier transforming is the most common approach to computing the frequency response of a time domain signal. When the signal is discrete in the time domain, we can use either the discrete Fourier transform (DFT) or the fast Fourier transform (FFT). The DFT can be used to calculate the frequency response at an arbitrary frequency that is computed via the formula:

$$\tilde{S}(f) = DFT\{S(n\Delta t)\} = \sum_{n=0}^{N} S(n\Delta t) e^{-j2\pi f n \Delta t} \Delta t \qquad (5.1)$$

Despite the advantage of the DFT that it is not limited to equispaced samples in the frequency domain, it is relatively slow and is not suitable for the situation where the number of frequencies at which the frequency response is derived is large. In contrast to the DFT, the FFT is much faster, but it assumes that the time signal is periodic, and it yields the frequency response, which is also discrete and periodic. To increase the sampling rate in the frequency domain, we need to zero-pad the time signal to increase its time span, whose inverse determines the sampling rate in the frequency domain. In addition, the normal FFT algorithms based on radix 2 require the length of the signal to be 2^N, otherwise the signal also needs to be zero-padded. The zero-padding requires that the signal be small enough at the end of the simulation, otherwise significant truncation errors may be introduced.

In practice, we usually decrease the cell size and the time step to reduce the numerical dispersion of the FDTD algorithm so the fields in both time and spatial domains are oversampled. Before taking the DFT or FFT, it is useful to reduce the sampling rate of the time domain signal to save on the simulation time. For example, for an excitation pulse shown in Figure 5.2, we define the frequency where the spectrum is 1/10 of the maximum value as the effective maximum frequency f_m, and the frequency where the spectrum is 1/1,000 of the maximum value as the maximum frequency f_{max}. Usually these two are different, but not very far from each other. To assure that the simulation is accurate, we choose the cell size to be 1/20 of the wavelength of the effective maximum frequency, or $\Delta = c/20 f_m$. The time step is then $\Delta t = \Delta/\sqrt{3} c = 1/20\sqrt{3} f_m$. On the other hand, according to the Nyquist sampling theorem [4], the sampling frequency needs to be greater than twice the maximum frequency, and the sampling rate is $\Delta t' \leq 1/2 f_{max}$.

The ratio between the two is $\Delta t'/\Delta t = 10\sqrt{3} f_m / f_{max}$. In other words, the frequency response would be sufficiently accurate even if we output the field from the FDTD every 14 to 15 time steps, rather than at every time step. In some problems, of course, the output quantities have broader spectrum than the input pulse. Conservatively speaking, outputting every tenth time step should be adequate for this case. In some cases, there exist very fine structures and the cell size is much smaller than the wavelength at the maximum frequency, and the sampling rate can be further reduced in this case. Not only is the efficiency of the Fourier transform improved when we do this, the memory requirement to store the time domain field output is reduced as well.

When resonant structures are present in the simulation region, the time domain signal usually converges very slowly. If we wait for the signal to fully converge, the total simulation time can be very long. This situation can be handled in two ways: the simple way is to apply a time window to the signal and force the signal to converge; the second approach is to extract the physical model of the resonance and use it to predict the signal beyond the end of the simulation period. In this section, we discuss the time windowing, and defer the topic of signal extrapolation to the next section.

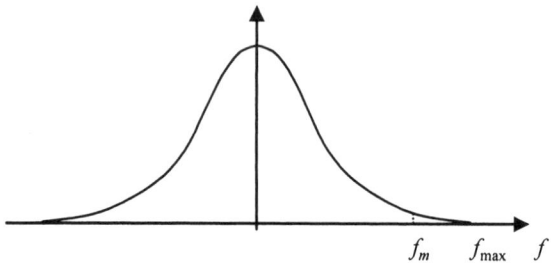

Figure 5.2 Spectrum of an excitation pulse.

Commonly used window functions in digital signal processing include Hamming, Hanning, Blackman, and Kaiser window [4]. These windows are symmetric, and they provide damping in both directions in time. In the FDTD, however, we know that the time signal is causal, and only need to use one-half of the window. In fact, this is also necessary to do, because otherwise the early time information will be distorted. Since the early time response usually contains the frequency response in most of the band, it is more reasonable to multiply this section of the signal with the smooth section of the time window.

In many applications, the selection of the source also helps to improve the convergence of the time signal. For example, we can: (1) change the source structure, for instance use coax to replace the probes to direct the energy away from the system, or (2) use a loaded and matched source to absorb the energy in the system.

5.2 MODE EXTRACTION OF TIME SIGNAL

When a signal is slowly converging, its late time behavior is typically comprised of harmonic resonance type of contributions that can be described by a simple mathematical model. If we skip a few cycles of such a resonance, the model parameters can be extracted and used to predict the signal at a future time, simply via extrapolation and without further FDTD simulation. This approach is referred to as the "super resolution" technique because the frequency resolution it yields is higher than that of the Fourier transform. Of course the extrapolation procedure should not only be mathematically sound but physically valid as well.

The late time signature can often be described as a damped harmonic resonance, which can be expressed as:

$$x(t) = Ae^{-\alpha t} e^{j2\pi f_0 t} \tag{5.2}$$

where f_0 is the resonant frequency and α is the damping factor. When the system has more than one resonance, the signal can be represented as:

$$x(t) = \sum_{k=1}^{K} A_k e^{j\tilde{\omega}_k t} \tag{5.3}$$

where $\tilde{\omega}_k = 2\pi f_k + j\alpha_k$ is the complex resonant frequency whose imaginary part is the damping factor. For a real signal such as the time signature derived from the FDTD, we can use complex conjugates resonant frequency to construct the model. Although (5.3) represents a signal with multiple resonances, it should not be used during the early time period until the signal has stabilized in a form that is compatible with (5.3). One way to assure that the extrapolation is reliable is to validate it by using only a part of the time signature to determine the relevant parameters in (5.3) and testing to see how well it predicts the signal at future times, which have also been computed directly with the FDTD. Good agreement between the two would indicate that the extrapolation is reliable and can be used with confidence for late time prediction.

There are several ways to extract the model coefficients in (5.3), such as Padé [5, 6], Prony [6, 7], and ESPRIT [8], and the mathematical approaches to be followed in implementing these methods are all different. Here we only discuss Prony's method, which we use frequently more often than the other two. In Prony's method we begin by writing (5.3) into a discrete form:

$$x(n) = \sum_{k=1}^{K} A_k \left(e^{j\tilde{\omega}_k \Delta t}\right)^n \tag{5.4}$$

and its z-transform as:

$$\tilde{x}(z) = \sum_{k=1}^{K} \frac{A_k}{1 - \left(e^{j\tilde{\omega}_k \Delta t}\right) z^{-1}} \tag{5.5}$$

where K is the number of poles and each term $e^{j\tilde{\omega}_k \Delta t}$ is a pole of $\tilde{x}(z)$. We then write (5.5) into a rational function form:

$$\tilde{x}(z) = \sum_{k=0}^{K-1} b_k z^{-1} \bigg/ \left(1 + \sum_{k=1}^{K} a_k z^{-1}\right) \qquad (5.6)$$

where b_k and a_k are unknown coefficients, and $e^{j\tilde{\omega}_k \Delta t}$ is the root of the polynomial in the denominator of (5.6). In the frequency domain, (5.6) can be rewritten as:

$$x(n) + \sum_{k=1}^{K} a_k x(n-k) = \sum_{k=0}^{K-1} b_k \delta(n-k) \qquad (5.7)$$

where the right-hand side is equal to zero when $n > K-1$. Prony's method utilizes the known values of $x(n)$ to solve (5.7) for a_k. The next step is to solve for the roots of the polynomial in the denominator of (5.6) to extract the complex extrapolation appearing in (5.4).

In (5.7), it is necessary to guarantee that $n > K$, so that the number of sampling points should be greater than $2K$. There are many ways to solve for a_k. The minimum mean square method is easier to apply, but is not easy to estimate the model order K; hence, the process needs to be tested several times with different orders. The other way is to use singular value decomposition (SVD), in which the order K can be determined by comparing the singular values and retaining only those above a certain threshold. The SVD is reliable and is frequently used in practice.

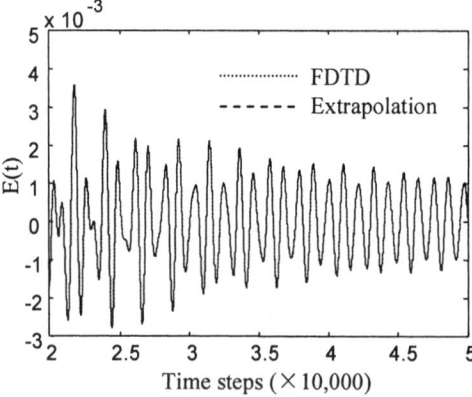

Figure 5.3 Model extraction using Prony's method.

In Figure 5.3, we note that the signal continues to resonate for a very long period of time and does not seem to have converged even at 50,000 time steps. To extrapolate this signal, we work with the signal from 20,000 to 25,000 time steps, reduce the sampling rate by 20 times, and then apply Prony's method. A fourteenth-order model is extracted from this process and used to predict the signal beyond the 25,000th time step. We can see that the predicted signal is almost identical to the simulated one. We should note that for (5.5) to be valid, the complex frequencies in (5.3) should have positive imaginary parts, which is also the condition for the system to be stable. If the extracted model does not satisfy this condition, the model is nonphysical and should not be used to predict the signal. In this case, we should either increase the model order or use the signal at a later time to extract the model coefficients, and ensure that they are indeed physical.

5.3 CIRCUIT PARAMETERS

Microwave network problems are very common in electromagnetic simulations. In such problems, we are interested in the network parameter matrices, including the S, Z, Y, T, and $ABCD$ matrices. These parameters are solved from the voltage and current outputs at the ports in the FDTD simulation.

To derive network parameter we utilize their definitions. For example, in an N-port network, shown in Figure 5.4, we can define the normalized voltage and current v_n and i_n for each of the ports. The incident and scattered waves at each port are given by $a_n = (v_n + i_n)/2$ and $b_n = (v_n - i_n)/2$, respectively. If all the ports are terminated by absorbing boundary conditions, then the incident waves at all ports are approximately zero with the exception of the excited port. Exciting one port at a time, we obtain one column of the S-parameter matrix by using:

$$S_{i,j} = \frac{b_i}{a_j}\bigg|_{a_k=0, k \neq j} \tag{5.8}$$

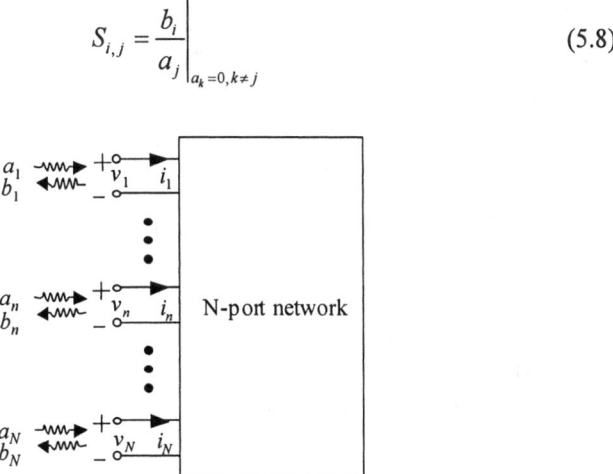

Figure 5.4 Determination of network parameters.

Obtaining the Z matrix using the same configuration is more difficult, because the currents at all ports are not zero, so that we cannot apply directly the definition of the Z-parameter matrix:

$$Z_{i,j} = \frac{v_i}{i_j}\bigg|_{i_k=0, k\neq j} \quad (5.9)$$

Instead, we need to explicitly write the equations relating the voltages, currents, and the Z matrix, and solve an element of this matrix for unknowns whose number of unknowns in this case is N^2, where N is the number of equations associated with each port. Thus we have to excite all the ports, one at a time, to derive the Z matrix.

For a uniform transmission line, we have discussed the mode-matching technique in Chapter 4 to calculate the amplitude of the wave at any point. If a port is not terminated by a matched load, we need to measure the wave amplitude at two or more points on the transmission line and solve for the amplitude of the forward and backward traveling waves. Suppose the transmission is in the z-direction, and we want to find out the voltage wave on the transmission line. At any point of the line voltage satisfies:

$$\tilde{V}(z) = \tilde{V}_+ e^{-j\beta z} + \tilde{V}_- e^{j\beta z} \quad (5.10)$$

where $\tilde{V}_+$ and $\tilde{V}_-$ are the traveling wave amplitude along the $+z$- and $-z$-directions, respectively, and β is the complex propagation constant. We sample the voltage at N points along the z-direction, and obtain the following matrix equation:

$$\begin{bmatrix} e^{-j\beta z_1} & e^{j\beta z_1} \\ \vdots & \vdots \\ e^{j\beta z_N} & e^{j\beta z_N} \end{bmatrix} \begin{bmatrix} \tilde{V}_+ \\ \tilde{V}_- \end{bmatrix} = \begin{bmatrix} \tilde{V}(z_1) \\ \vdots \\ \tilde{V}(z_N) \end{bmatrix} \quad (5.11)$$

When the propagation constant is known, this overdetermined equation can be solved using the minimum mean square method. Otherwise, we can also use Prony's method to determine the propagation.

Usually, the propagation constant is not known, and we rewrite (5.10) in a more general form without forcing the propagation constant to be the same except for the sign in both directions:

$$\tilde{V}(z) = \tilde{V}_+ e^{-j\tilde{K}_+ z} + \tilde{V}_- e^{j\tilde{K}_- z} \tag{5.12}$$

In (5.12) there are 4 unknowns $\tilde{V}_+$, $\tilde{K}_+$, $\tilde{V}_-$, and $\tilde{K}_-$. Our goal is to solve for them using the voltages measured at a minimum of four points uniformly sampled along the line. So we have:

$$\tilde{V}_{z_n} = \sum_{m=1}^{2} \tilde{V}_m e^{j\tilde{K}_m z_n} \tag{5.13}$$

where $\tilde{V}_1 = \tilde{V}_+$, $\tilde{K}_1 = \tilde{K}_+$, $\tilde{V}_2 = \tilde{V}_-$, and $\tilde{K}_2 = \tilde{K}_-$. To apply Prony's method [7, 9], let $\tilde{x}_m = e^{j\tilde{K}_m z_n}$ in the above equation and it becomes:

$$\tilde{V}_{z_n} = \sum_{m=1}^{2} \tilde{V}_m \tilde{x}_m \tag{5.14}$$

where $\tilde{x}_m$ are the roots of the following polynomial equation:

$$\alpha_2 + \alpha_1 \tilde{x}_m - \tilde{x}_m^2 = 0 \tag{5.15}$$

where α_1 and α_2 are the coefficients that are yet to be determined. Next, we multiply the first three equations in (5.14) with α_2, α_1, and -1 and add them together, and then do the same with the last three equations and add them together, then by applying the condition (5.15), we obtain the following equations:

$$\alpha_2 \tilde{V}_{x_1} + \alpha_1 \tilde{V}_{x_2} - \tilde{V}_{x_3} = 0 \tag{5.16a}$$

$$\alpha_2 \tilde{V}_{x_2} + \alpha_1 \tilde{V}_{x_3} - \tilde{V}_{x_4} = 0 \tag{5.16b}$$

From (5.16) we can solve for α_1 and α_2, and then solve for $\tilde{x}$ from (5.15). Finally, we find $\tilde{K}_m$ from its definition $\tilde{x}_m = e^{j\tilde{K}_m z_n}$ to get:

$$\tilde{K}_m = \frac{1}{jz_n} \ln(\tilde{x}_m) \tag{5.17}$$

Substituting (5.17) into (5.12), we can again use the minimum mean square method to solve for $\tilde{V}_m$. This method usually works better at high frequencies, because (5.14) might become ill-conditioned at low frequencies. Another point to keep in mind is that we assumed single-mode transmission while developing the above procedure, but this may not be true in practice. So we either need to take the sampling points far away from any discontinuity to decrease the effect of higher-order modes, or increase the number of sampling points and extract the higher-order modes as well in (5.14).

5.4 NEAR-TO-FAR-FIELD TRANSFORMATION

Although the FDTD is a generalized tool for electromagnetic simulation, we do not use it to compute the far field directly because that would require working with such a very large computational domain as to render it impractical. Fortunately, the far field can be obtained through the spatial Fourier transform of the equivalent currents that are derived from the near fields. In this section, we briefly introduce the basic concept and acceleration methods of the near-to-far-field transformation procedure.

5.4.1 Basic Approach

According to the equivalence principle [10], it is possible to replace the sources enclosed by a closed surface with the equivalent currents on this surface, which reproduce the original fields in the region external to the surface. For example, if the source is located inside the surface S, as shown in Figure 5.5, we can configure an equivalent problem in which the field outside S is exactly the same as the original problem, and is zero inside S (see Figure 5.5(b)).

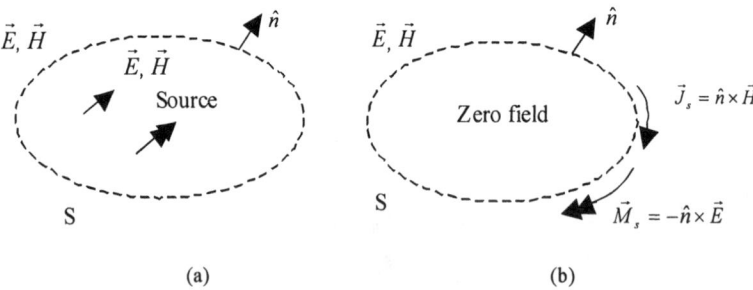

Figure 5.5 Illustration of the equivalence principle. (a) Original problem. (b) Equivalent problem.

To support the original field distribution in the exterior region, the equivalent current distribution on S should be chosen as:

$$\vec{J}_s = \hat{n} \times \vec{H} \qquad (5.18\text{a})$$

$$\vec{M}_s = -\hat{n} \times \vec{E} \qquad (5.18\text{b})$$

The surface S is usually referred to as the Huygens' surface. In the far field calculation, it should include all the sources and objects. To reduce the size of the computational domain, it is better to place the Huygens' surface as close to the objects as possible. Once we derive the field distribution on the Huygens' surface, we can compute the equivalent current sources and then the far field from these sources.

In the FDTD simulation, the sampling points of the electric and magnetic fields are not collated but are one-half cell apart. There are two ways to define the Huygens' surface [11–17] to resolve this problem. In the first approach, the Huygens' surface coincides with the interface on which either the tangential electric or the magnetic field is sampled in the FDTD. If the surface is that surface of the tangential electric field, the magnetic field needs to be interpolated on the Huygens' surface. Such interpolation may introduce numerical errors because the interpolated field is no longer second-order accurate. In the second approach, we define two Huygens' surfaces, one overlapping with the tangential electric field, and another one with the tangential magnetic field [12]. Next, we describe the details of the first approach to predicting the far field in the FDTD simulations. Although the second approach employs two separate Huygens' surfaces, the procedure to compute far field is no difference.

Suppose our Huygens' surface includes all the objects and antennas, and an observation point is placed in the far zone. The field at the cell center on the Huygens' surface, shown in Figure 5.6, can be obtained by using a weighted average of the neighboring fields. We can then derive the equivalent electric and magnetic currents from (5.18).

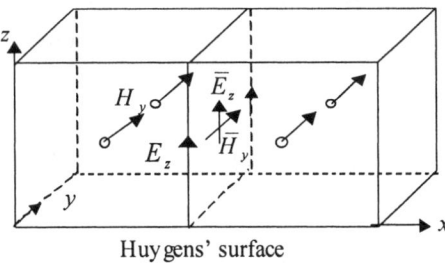

Huygens' surface

Figure 5.6 Equivalent fields on the Huygens' surface.

For example, the fields $\bar{H}_y^{n-1/2}(i+1/2, j, k+1/2)$ and $\bar{E}_z^n(i, j, k+1/2)$ can be written as:

$$\bar{H}_y^{n-1/2}(i+1/2, j, k+1/2) =$$
$$\frac{0.5\Delta x(i-1)}{\Delta x(i)+\Delta x(i-1)} \frac{H_y^{n-1/2}(i-1/2, j, k+1/2) + H_y^{n-1/2}(i-1/2, j+1, k+1/2)}{2} \quad (5.19)$$
$$+ \frac{0.5\Delta x(i)}{\Delta x(i)+\Delta x(i-1)} \frac{H_y^{n-1/2}(i+1/2, j, k+1/2) + H_y^{n-1/2}(i+1/2, j+1, k+1/2)}{2}$$

$$\bar{E}_z^n(i, j, k+1/2) = \frac{E_z^n(i, j, k+1/2) + E_z^n(i, j+1, k+1/2)}{2} \quad (5.20)$$

and the resulting equivalent currents will be:

$$M_y^n(i, j, k) = -\hat{n}_x \times \vec{E}_z^n(i, j, k) \quad (5.21)$$

$$J_z^{n-1/2}(i, j, k) = \hat{n}_x \times \vec{H}_y^{n-1/2}(i, j, k) \quad (5.22)$$

where $\hat{n}_x$ is the unit vector in the x-direction. The far field can then be obtained from the following formulations [18]:

$$E_\theta = -\frac{j\beta \exp(-j\beta r)}{4\pi r}(L_\varphi + \eta N_\theta) \quad (5.23)$$

$$E_\varphi = \frac{j\beta \exp(-j\beta r)}{4\pi r}(L_\theta - \eta N_\varphi) \quad (5.24)$$

where $\eta = \sqrt{\mu_0/\varepsilon_0}$ is the wave impedance in free space and $\beta = \omega\sqrt{\mu_0 \varepsilon_0}$ is the propagation constant in free space. In the Cartesian coordinate system, the terms N_θ, N_φ, L_θ, and L_φ can be written as:

$$N_\theta = \iint_S \left[J_x \cos\theta \cos(\varphi) + J_y \cos\theta \sin(\varphi) - J_z \sin\theta\right] \exp(j\beta r' \cos\varphi) ds' \quad (5.25)$$

$$N_\varphi = \iint_S \left[-J_x \sin(\varphi) + J_y \cos(\varphi)\right] \exp(j\beta r' \cos\varphi) ds' \quad (5.26)$$

$$L_\theta = \iint_S \left[M_x \cos\theta\cos(\varphi) + M_y \cos\theta\sin(\varphi) - M_z \sin\theta\right]\exp(j\beta r'\cos\varphi)ds' \quad (5.27)$$

$$L_\varphi = \iint_S \left[-M_x \sin(\varphi) + M_y \cos(\varphi)\right]\exp(j\beta r'\cos\varphi)ds' \quad (5.28)$$

where J_x, J_y, J_z, M_x, M_y, and M_z are the electric and magnetic current densities on the surface of the Huygens' box. In (5.25) through (5.28), the angles φ and θ are the observation angles, and x', y', and z' are the coordinates associated with the currents. The variable ds' can be approximated by $dx' \times dy'$, $dy' \times dz'$, and $dx' \times dz'$, respectively, in the x-y, y-z, and x-z planes. Also $r'\cos\phi$ can be expressed as $(x'\cos\theta\cos\varphi + y'\sin\theta\sin\varphi)$ in the x-y plane, $(y'\sin\theta\sin\varphi + z'\cos\theta)$ in the y-z plane, and $(x'\sin\theta\cos\varphi + z'\cos\theta)$ in the x-z plane.

5.4.2 Far-Field Calculation for Planar Structures

Frequently, while simulating planar radiators such as microstrip patch antennas, we model the ground plane and dielectric substrate to be infinitely large if they are much wider than the patch. The use of approximation not only enables us to reduce the computational domain considerably but also to match the source more easily. For a finite structure, the far field is obtained from the Huygens' surface that encloses the structure in its entirety. However, it is not possible to do this when the problem is infinitely large, and we can only choose the Huygens' surface to be an infinite plane above the patch antenna. Since the computational domain must be finite, we truncate the Huygens' surface at the sides by the domain boundaries. Consequently, in order to obtain an accurate solution, we have to use a larger domain, so that the Huygens' surface is also large. Usually, due to the existence of the ground plane, the tangential electric field on the Huygens' surface parallel to the plane decays much faster than the tangential magnetic field as we move away from the object. Using the image theorem and the equivalence principle, we can derive the far field using only the magnetic current, by placing an artificial PEC one half space below the Huygens' surface [10]. Numerical results show that the size of the Huygens' surface can be considerably reduced to one that is only slightly larger than the patch. We demonstrate this by using a simple example.

Let us consider a simple patch antenna configuration shown in Figure 5.7. We first choose a large computational domain size whose transverse dimensions are 380×380 mm. The Huygens' surface is placed two cells above the patch, and its size is the same as that of the domain in the horizontal direction. The electric and magnetic field distributions on the Huygens' surface at the frequency of 1.8 GHz are plotted in Figure 5.8. It is evident that the electric field (magnetic current) is more localized than the magnetic field (electric current), and the latter decays very slowly.

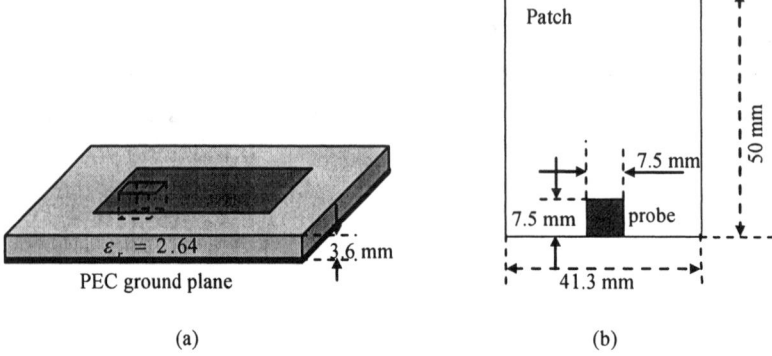

Figure 5.7 Configuration of a patch antenna. (a) Antenna configuration. (b) Patch dimension.

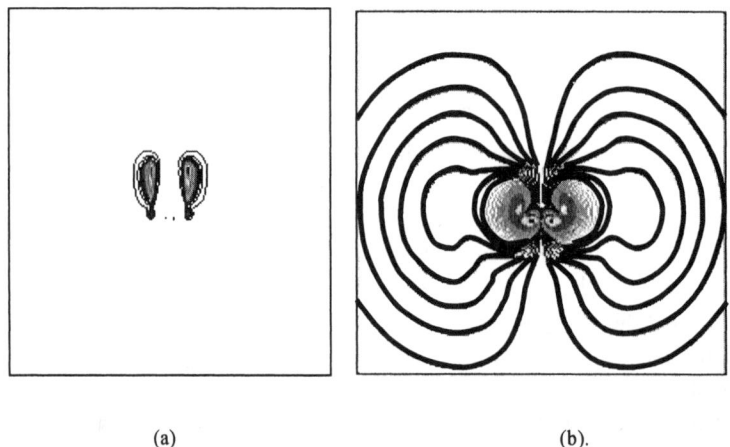

Figure 5.8 Electric and magnetic field distribution on the Huygens' box at f = 1.76 GHz. (a) Electric field distribution. (b) Magnetic field distribution.

To get a quantitative estimate of the field behaviors, we choose the cross sections in the x- and y-directions in Figure 5.8(b), in which the magnetic field reaches the maximum value, and plot the field distribution on the cross section in Figure 5.9. The antenna size is shown as the dashed line. We can see that even if we choose a very large Huygens' surface, there will still be truncation errors at the domain boundaries, which is undesirable.

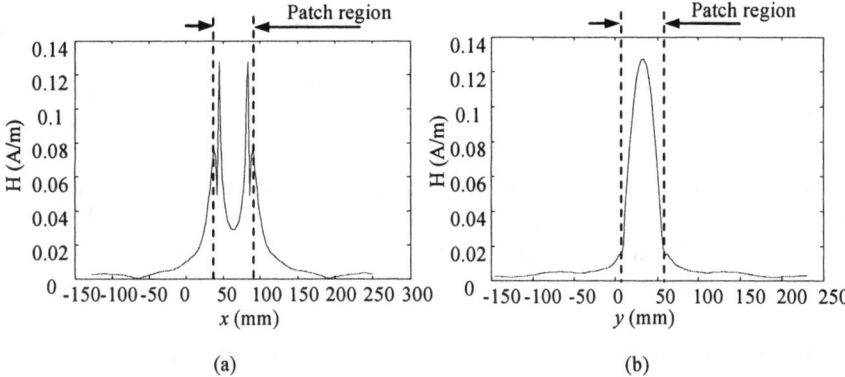

Figure 5.9 Magnetic field distribution in the x and y cross sections. (a) Magnetic field along the x-direction. (b) Magnetic field along the y-direction.

Next, we study the behavior of the electric field (magnetic current) on the cross sections and plot it in Figure 5.10.

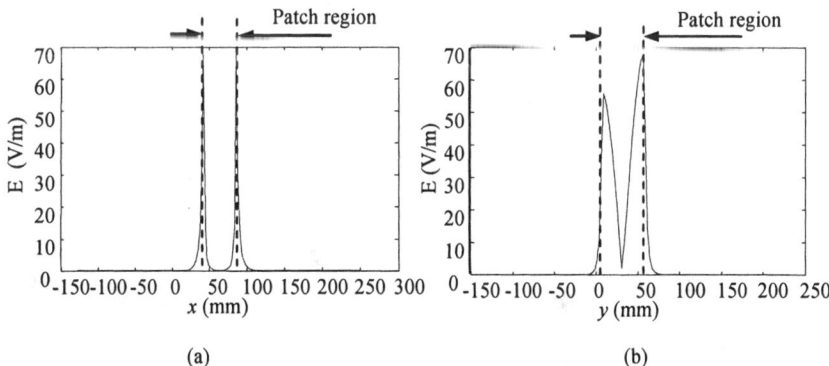

Figure 5.10 Electric field distribution in the x and y cross sections. (a) Electric field along the x-direction. (b) Electric field along the y-direction.

From Figures 5.9 and 5.10 we can see that in contrast to the magnetic field, the electric field on the Huygens' surface can be neglected when the observation is only slightly away from the patch. Thus, we use only the magnetic current to calculate the far field, and the results would be very accurate even if the Huygens' surface is just slightly wider than the patch.

In this example, we are also interested in the directivity of the antenna, and this requires the calculation of the total radiated power, which can be obtained by integrating the Poynting vector in the far zone. However, the drawback of this approach is that we have to calculate the radiated fields in all directions. For an electrically large antenna or antenna array, this is very time consuming. A simpler way is to integrate the Poynting vector on the Huygens' surface. By doing this, we circumvent the need to compute the radiated power via the far field and thereby reduce the computational time substantially.

To verify the approach above, we first calculate the directivity of the patch antenna using both the electric and magnetic currents, and then repeat the same calculation with the magnetic current only. The results are summarized in Figure 5.11. As a reference check, the same problem is simulated with a commercial method of moment (MoM) software and the result is plotted in the same figure. It can be seen that when both the electric and magnetic currents are used, the result converges to the correct answer as the Huygens' surface gets larger and larger. But even when the Huygens' surface is 12 times larger than the magnetic current, the solution is still not sufficiently accurate. We should also mention that the incremental changes in the directivity are not noticeable when we increase the size of the Huygens' box if we use the magnetic current only.

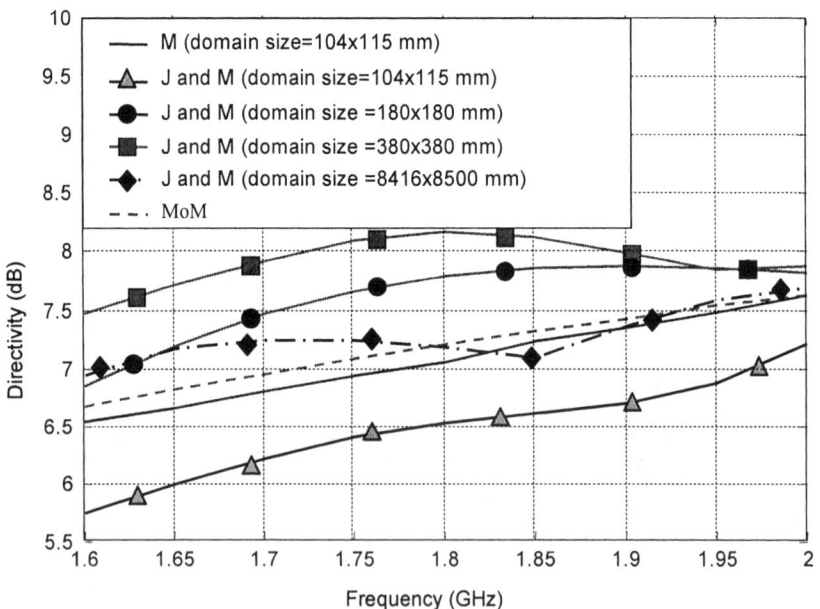

Figure 5.11 Directivity of the patch antenna calculated using different approaches.

5.4.3 Data Compression and Its Application

In this section, we discuss the use of data compression [19] in the near-to-far field transformation. There are two commonly used approaches for computing the far fields in the FDTD. The first one of these specifies the frequency sampling points before the simulation, and uses the discrete Fourier transform (DFT) to compute the frequency domain currents on the Huygens' box at each time step. The advantage is that the required storage is relatively small and we do not have to output any intermediate files. However, the disadvantage is that if we want to change the observation frequencies, we have to start the simulation anew. If we use the fast Fourier transform (FFT) at the end of the simulation, then we would need to store the time domain field at each point and every time step on the Huygens' box. The other approach is the time domain near-to-far-field transformation [13], in which the time domain far field is calculated for every observation angle and at each time step. This is the preferred choice when the number of observation angles is small, as it is, for instance, in a monostatic scattering problem.

We have discussed time domain compression schemes at the beginning of this chapter, which enable us to reduce the sampling rate greatly. In fact, the field on the Huygens' surface is typically oversampled beyond what is needed for accurate prediction of the far fields. Based on the numerical dispersion of the FDTD grid, the cell size is at most one-tenth of the minimum wavelength. In practice, however, cell sizes as small as one-hundredth or even one-thousandth of a wavelength are sometimes used, in order to accurately model objects with the fine features. For the far-field calculation, however, only a few cells per wavelength are adequate. A simple way to reduce the burden of computation is to apply a lowpass filtering of the field on the Huygens' surface, by using a larger cell size, and take the field as the average of these computed on the actual FDTD grid [18]. After down-sampling in both the time and spatial domains, the data storage can be compressed to a manageable range. It should be pointed out that the down-sampling technique in the time and spatial domains are not the same. In the time domain, we simply take the field value once every few time steps. But in the spatial domain, the field is only oversampled in terms of the visible range of the far field, but not always for the near field. For this reason we need to apply lowpass filtering before down-sampling the field. Next, we introduce the data compression technique and its application to the near-to-far-field transformation.

Data compression, which is an important branch in signal processing, is usually categorized into lossless compression or lossy compression schemes [20, 21]. The discussion of this topic is beyond the scope of this book, and we only introduce the discrete cosine transform (DCT) as an implementation of the low-loss compression technique. The formulation of the DCT is as follows:

$$c(u,v) = \frac{4}{MN} E(u)E(v) \sum_{i=0}^{M-1} \sum_{j=0}^{N-1} f(i,j) \cos\left[\frac{(2i+1)}{2M} u\pi\right] \cos\left[\frac{(2j+1)}{2N} v\pi\right] \quad (5.29)$$

where M and N are grid numbers in the x- and y-directions, respectively, and $u = 0, 1,..., M-1$, $v = 0, 1,..., N-1$. For $u, v = 0$, we choose both $E(u)$ and $E(v)$ to be equal to $1/\sqrt{2}$, otherwise, both $E(u)$ and $E(v)$ should be chosen to be equal to 1. Also, $f(i,j)$ is the original current distribution on the Huygens' surface.

The original current distribution covers the entire Huygens' surface. After the transformation given in (5.29), the field transformed into the spectral domain and its distribution $c(u,v)$ is more concentrated near the u- and v-axes. For the purpose of the far-field calculation, we only need to retain a small portion of the data. The actual amount of storage depends on the frequency range of interest and the accuracy decreased. To calculate the far field, we first use the inverse discrete cosine transform (IDCT) to reconstruct the current distribution on the Huygens' surface, and then carry out its near-to-far-field transformation. The IDCT is expressed as:

$$f(i,j) = \sum_{u=0}^{M-1} \sum_{v=0}^{N-1} E(u)E(v)c(u,v) \cos\left[\frac{(2i+1)}{2M} u\pi\right] \cos\left[\frac{(2j+1)}{2N} v\pi\right] \quad (5.30)$$

where $u = 0, 1,..., M-1$, $v = 0, 1,..., N-1$. For $u, v = 0$, we choose both $E(u)$ and $E(v)$ to be equal to $1/\sqrt{2}$; otherwise, both $E(u)$ and $E(v)$ are equal to 1.

Next, we demonstrate the application of the DCT technique through a simple example, namely, a rectangular patch antenna. The ground plane and the dielectric substrate of the patch antenna are infinite in the horizontal directions. The Huygens' surface is located two cells above the antenna, and has the same size as that of the top surface of the computational domain. The domain size is 23.34×40 × 3.71 mm, and it is discretized into 60 × 100 × 14 uniform cells. The output in which we are interested is the far field.

A pure Gaussian source is placed between the microstrip feed and the ground, and the highest frequency in its spectrum is 27 GHz. The walls of the computational domain with the exception of the bottom one are truncated by PML.

We plot the E_x field distribution on the Huygens' surface at the 400th time step in Figure 5.12. Using the DCT (5.29), we can obtain the transformed field as shown in Figure 5.13. The time domain down-sampling used is 1 point every 35 time steps, determined by the maximum frequency of the source field.

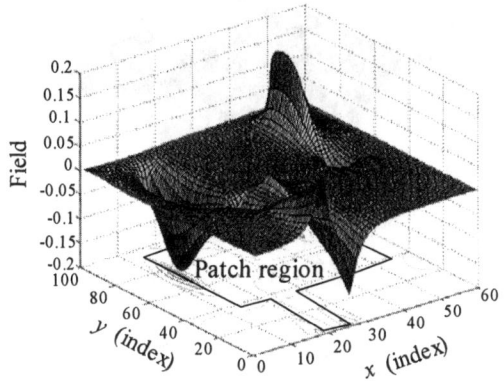

Figure 5.12 Electric field distribution at time step 400.

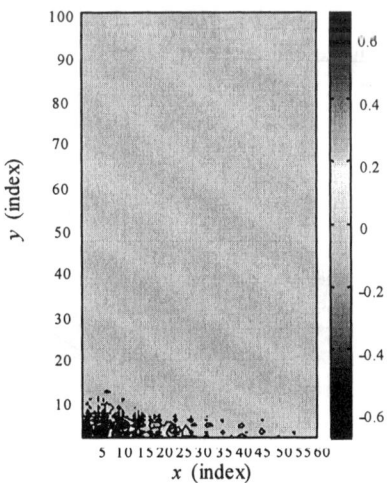

Figure 5.13 Electric field distribution in the spectrum domain at time step 400.

The data we store into the data file is only 7.5% in the above figure, and the rest can be discarded. This data is used to reconstruct the field distribution before calculating the far field. The reconstruction is performed using the IDCT, and the reconstructed field is shown in Figure 5.14.

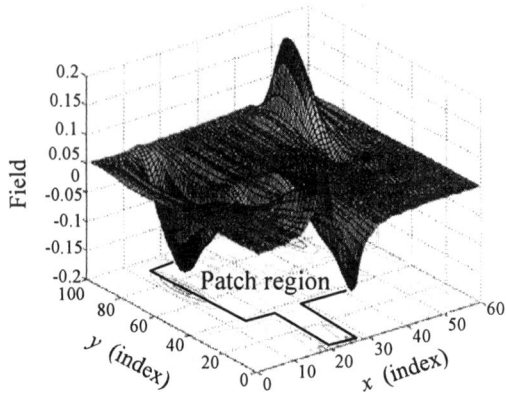

Figure 5.14 Electric field distribution reconstructed by the IDCT.

In this example, we only store the information in a 30×15 region (7.5% of the original size) for the far-field calculation. Actually for frequencies below 10 GHz, a 3×3 local region is adequate for accurate computation of the far field. We compute the far field at 5 GHz using the procedure described above, and compare it with that obtained from the original data in Figure 5.15. The agreement between the two is seen to be very good. In order to increase the data compression rate and its compression efficiency, further research into the transformations (5.29) and (5.30) is necessary.

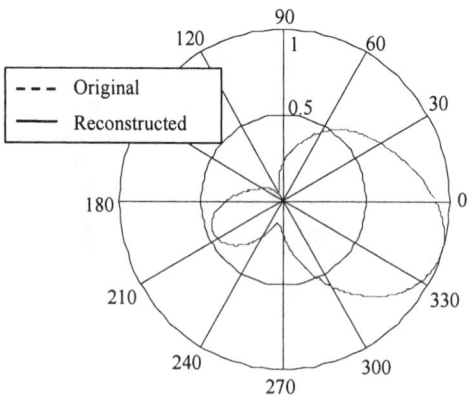

Figure 5.15 Far field obtained using the original and the reconstructed near fields.

REFERENCES

[1] W. Gwarek and M. Celuch-Marcysiak, "Wide-Band S-Parameter Extraction from FD-TD Simulations for Propagating and Evanescent Modes in Inhomogeneous Guides," *IEEE Transactions on Microwave Theory and Techniques*, Vol. 51, No. 8, August 2003, pp. 1920-1928.

[2] Y. Wang and H. Ling, "Multimode Parameter Extraction for Multiconductor Transmission Lines Via Single-Pass FDTD and Signal-Processing Techniques," *IEEE Transactions on Microwave Theory and Techniques*, Vol. 46, No. 1, January 1998, pp. 89-96.

[3] T. Huang, B. Houshmand, and T. Itoh, "Efficient Modes Extraction and Numerically Exact Matched Sources for a Homogeneous Waveguide Cross-Section in a FDTD Simulation," *IEEE MTT-S Int. Microwave Symp.*, San Diego, CA, Vol. 1, May 1994, pp. 31-34.

[4] S. Mittra, *Digital Signal Processing*, McGraw-Hill, New York, 1998.

[5] S. Dey and R. Mittra, "Efficient Computation of Resonant Frequencies and Quality Factors of Cavities Via a Combination of the Finite-Difference Time-Domain Technique and the Padé Approximation," *IEEE Microwave and Guided Wave Letters*, Vol. 8, No. 12, December 1998, pp. 415-417.

[6] J. Ma, W. Yu, and R. Mittra, "Detection of Buried Dielectric Cavities Using the Finite-Difference Time-Domain Method in Conjunction with Signal Processing Techniques," *IEEE Transactions on Antennas and Propagation*, Vol. 48, No. 9, September 2000, pp. 1289-1294.

[7] M. Hayes, *Statistical Digital Signal Processing and Modeling*, John Wiley & Sons, New York, 1996.

[8] R. Roy, A. Paulraj, and T. Kailath, "ESPRIT — A Subspace Rotation Approach to Estimation of Parameters of Cisoids in Noise," *IEEE Transactions on Acoustics, Speech, and Signal Processing*, Vol. 34, October 1986, pp. 1340-1342.

[9] C. Pearson and D. Roberson, "The Extraction of the Singularity Expansion Description of a Scatterer from Sampled Transient Surface Current Response," *IEEE Transactions on Antennas and Propagation*, Vol. 28, No. 1, 1980, pp. 182-190.

[10] R. Harrington, *Time-Harmonic Electromagnetic Fields*, Wiley-IEEE Press, New York, 2001.

[11] R. Luebbers, et al., "A Finite Difference Time Domain Near-Zone to Far-Zone Transformation," *IEEE Transactions on Antennas and Propagation*, Vol. 39, No. 4, April 1991, pp. 429-433.

[12] T. Martin, "An Improved Near-to-Far Zone Transformation for the Finite-Difference Time-Domain Method," *IEEE Transactions on Antennas and Propagation*, Vol. 46, No. 9, September 1998, pp. 1263-1271.

[13] K. Kunz and R. Luebbers, *The Finite-Difference Time Domain Methods for Electromagnetics*, CRC Press, Boca Raton, FL, 1993.

[14] S. Shum and K. Luk, "An Efficient FDTD Near-to-Far-Field Transformation for Radiation Pattern Calculation," *Microwave and Optical Technology Letters*, Vol. 20, No. 2, 1999, pp. 129-131.

[15] B. Randhawa, J. Tealby, and A. Marvin, "Modification to Time Domain Near-Field to Far-Field Transformation for FDTD Method," *Electronics Letters*, Vol. 313, No. 25, December 1997, pp. 2132-2133.

[16] Z. Huang and R. Plumb, "An FDTD Near- to Far-Zone Transformation for Scatterers Buried in Stratified Grounds," *IEEE Transactions on Antennas and Propagation*, Vol. 44, No. 8, August 1996, pp. 1150-1157.

[17] R. Luebbers, D. Ryan, and J. Beggs, "A Two-Dimensional Time-Domain Near-Zone to Far-Zone Transformation," *IEEE Transactions on Antennas and Propagation*, Vol. 40, No. 7, July 1992, pp. 848-851.

[18] C. Balanis, *Advanced Engineering Electromagnetics*, John Wiley & Sons, New York, 1989, pp. 614-618.

[19] B. Zhou, S. Wang, and W. Yu, "An Effective Approach to Predict Far Field Pattern in FDTD Simulation," *IEEE Antenna and Wireless Propagation Letters*, Vol. 3, No. 9, 2004, pp. 148-151.

[20] M. Nelson and J. Gailly, *The Data Compression Book*, 2nd ed., M&T Books, New York, 1995.

[21] P. Duhamel and C. Guillemot, "Polynomial Transform Computation of the 2–D DCT," *IEEE Proc. International Conference Acoustic, Speech and Signal Processing*, Piscataway, NJ, 1990, pp. 1515-1518.

Chapter 6

Introduction to Parallel Computing Systems

Flynn's taxonomy classifies machines according to whether they have single or multiple streams [1]. The four combinations are the single instruction stream together with single data stream (SISD), the single instruction stream and multiple data streams (SIMD), multiple instruction streams and single data stream (MISD), and multiple instruction streams and multiple data streams (MIMD). Conventional computers with a single processor are classified as the SISD systems, in which each arithmetic instruction initiates an operation on a data item taken from a single stream of data elements. Historical supercomputers such as the Control Data Corporation 6600 and 7600 fit this category, as do most contemporary microprocessors. The SIMD machines have one instruction processing unit and several data processing units. There are few machines in the MISD category, and none of them has been commercially successful. The category of the MIMD machines is the most diverse of the four classifications in Flynn's taxonomy and most computers today are built using the MIMD technique. Moreover, the MIMD system includes four types of computers: the symmetric multiprocessor (SMP), the massively parallel processor (MPP), the distributed shared memory (DSM), and the cluster of workstation (COW).

Historically, there were four stages during the development of parallel machines. The first stage began in the late 1960s, and a representative machine was ILLIAC IV, which was a pioneer in massively parallel computing. It had 64 processing elements working together, and was designed to have up to 256. All the processing elements executed the same instruction simultaneously, and each one had its own local memory. The computer was the fourth in a line of computers built at the University of Illinois starting in 1948. It was moved to the NASA Ames Research Center in 1974, where its applications included fluid dynamics, digital image processing, and weather forecasting. However, ILLIAC IV was too expensive and had many technical problems. Only one was ever built, and it was finally retired in 1982. Cray supercomputers were the next generation of ILLIAC machines, and the Cray-1, announced in 1976, had a top speed of 133 megaflops. The first Cray-1 system was installed at the Los Alamos National Laboratory, and at least 16 Cray-1s were eventually built. A typical Cray-1 cost about $700,000 in 1976. The first one claimed a world record speed of 133 million floating-point operations per second (133 megaflops) and an 8-megabyte main memory. It achieved record speeds for various numerical calculations reaching a peak of 200 million instructions per second and 1 gigabit per second I/O transfer rate. The second stage was the development of PVP machines, which started in the 1970s, and the representative machines in this category were Cray YMP-90, NEC SX-3, and Fujitsu VP-2000. The parallel machines in the third stage were based on the shared memory system, and the representative machines were the SGI Challenge and Sun Sparc Center 2000. The fourth stage was the development of MPP machines starting from the late 1980s and the representative ones in this category were Intel Paragon, CM-5E, Cray T3D, and IBM SP2. "Constructing a supercomputer with personal computers" became the focus of the next series of developments that followed and it, in fact, became reality in the 1990s, as a large number of personal computers (PCs) with distributed memories were connected with each other via a network. Using this strategy, it has now become possible to fabricate high-performance computers that are comparable to supercomputers, but are orders of magnitude lower cost.

As mentioned previously, the MPI is an international standard that supports message passing in a parallel processing system, and provides a standard environment for this purpose as well as a standard communication library [2–4]. The MPI library is implemented through several implementation approaches such as the MPICH, LAM, Cray MPIProduct, IBM's MPI, and SGI's MPI. The MPICH, the most widely used implementation of the MPI, has been broadly used in variety of parallel computations.

In order to help the reader understand the MPI library and MPICH, in this chapter we first briefly introduce the architectures of parallel computing systems based on the MIMD technique. Next, we discuss some basic concepts of the MPICH architecture and parallel programming techniques. Finally, we analyze the efficiency of the parallel FDTD code when run on different platforms.

6.1 ARCHITECTURE OF THE PARALLEL SYSTEM

"Cluster" is a widely used terminology, and a computer cluster comprises a set of independent computers that are connected together into a unified system via special software and networking arrangement. Clusters are typically used for high availability (HA) for greater reliability or high performance computing (HPC) to provide greater computational power than a single computer can deliver. Here we focus on the high-performance computing system comprising a cluster. With development of the MIMD systems, the software environment supporting parallel computing techniques such as operating system, compiler, and parallel libraries have been addressed. In this section, we discuss the architecture of computers belonging to the MIMD system, namely, SMP, DSP, MPP, and Beowulf PC clusters.

6.1.1 Symmetric Multiprocessor

The basic unit of the SMP system [5, 6] is one of the commonly found microprocessors (e.g., the IBM PowerPC or the Intel Pentium4). The processors in the parallel computing system are connected to the system memory as well as output devices via its system data bus, switch, or HUB. The SMP system architecture is shown in Figure 6.1 in a block diagram.

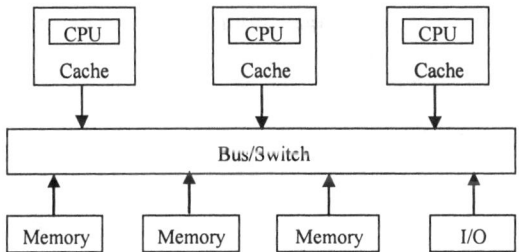

Figure 6.1 Architecture of the SMP systems.

The SMP system achieves high performance by making multiple CPUs available simultaneously to the individual process. Unlike asymmetrical processing, any idle processor in the SMP system can be assigned to any task, and additional CPUs can be added to improve the system performance and to handle the increased loads. All processors share the global memory in a "shared memory model." They also share the I/O subsystem.

The SMP processors run a single copy of the operating system. There is no master and slave relationship between the SMP processors, and they are all assigned equal positions. All of the processors scan the table of operating system processes and look for the highest priority process that is ready to run, and then they run the process when they find it. The processor returns to scanning the table of operating system processes when a process is blocked or finished. An SMP system may contain a fixed number of processors or it may accommodate upgrading the number and type of processors.

In order to overcome the bottleneck of memory access, the processors in the SMP system are not directly connected to the system memory, but instead to the memory via a high-speed cache. Generally speaking, the memory access is much slower than the processor speed, and even a single-processor machine tends to spend a considerable amount of time waiting for data to arrive from the system memory. The SMP system makes this situation even worse, as only one processor can access memory at a time; and it is possible that several processors could starve for the data to arrive. Therefore, the number of processors in the SMP system cannot usually be too large.

6.1.2 Distributed Shared Memory System

The architecture of a distributed shared memory (DSM) system is an extension of the SMP system [7]. However, its basic unit is a node; each unit usually includes multiple processors, local storage units, and HUBs; and each processor has a high-speed cache. The processors, local storage, and I/O devices are connected to each other via the HUB. The HUBs are connected to routers, and the routers of nodes are connected together and form a high-performance network of a parallel computing system. A typical structure of the DSM is shown in Figure 6.2.

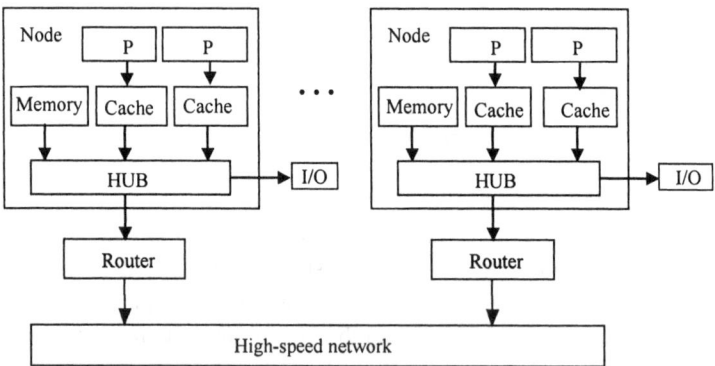

Figure 6.2 Architecture of the DSM system.

In a distributed shared memory system, each node is an independent computer system that has, at least, a processor and memory, and the nodes are connected together via a network. This is the most cost-effective way to construct a distributed shared memory system, because we can utilize many different types of workstations available, and connect them via a network without having to add any new hardware beyond what exists already. However, this strategy is sometimes ineffective for heavy computation, because, for instance, networks for the general purpose are slow, and the nodes may be unexpectedly used for other work, and hence it is difficult to schedule them efficiently.

The distributed shared memory systems have no memory bus problem. Each processor can use the full bandwidth to its own local memory without interference with other processors. Thus, there is no inherent limit to the number of processors. The size of the system is constrained only by the network used to connect the node computers. Some distributed shared memory systems consist of many thousand processors.

Since the nodes in a distributed shared memory system do not share the memory, data exchange among processors is more difficult in such a system than that in a shared memory system. A distributed shared memory system utilizes the message passing programming model, which is organized as a set of independent tasks that communicate with each other via the message passing protocol. This introduces two sources of overhead: it takes time to construct and send a message from one processor to another, and the receive processor must be interrupted to deal with the messages from other processors.

6.1.3 Massively Parallel Processing

Massively parallel processing (MPP) systems [8] are divided into two categories, according to the difference in storage structures. There are distributed memory massively parallel systems, and clusters constructed from the SMP and DSM systems by a high-performance network. A typical architecture of the MPP system is shown in Figure 6.3.

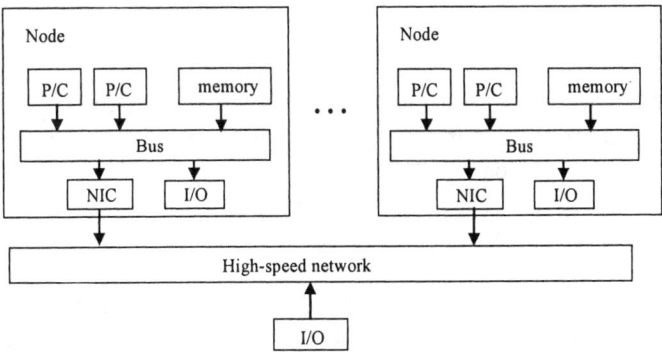

Figure 6.3 Architecture of the MPP system.

A massively parallel processor generally denotes a distributed memory computer with more than a "few" independent processors or nodes, perhaps hundreds or even thousands. In the MPP system, each node may be a single processor or a collection of the SMP processors. An MPP collection of the SMP nodes is sometimes called an SMP cluster. Each node has its own copy of the operating system, memory, and disk storage, and a particular mechanism of data or message passing makes each computer work on a different part of a problem. The software must be particularly programmed to take advantage of this type of architecture.

The MPP system is designed to obtain performance improvement through the use of large numbers (tens, hundreds or thousands) of individually simple, lowowered processors, and each processor has its own memory. The MPP system normally runs on the Unix system and must have more than one Unix kernel or more than one "ready to run" queue.

Distributed memory massively parallel processor (MPP) systems based on commodity chips have become increasingly important to supercomputing because of the low price performance ratios and extremely large memory available. For suitable problems, in particular for very memory-intensive problems, these systems are capable of much higher computational performance, and can compete in grand challenges and capability computing problems beyond the scope of other systems.

Unlike the SMP or DSM system, the nodes of the MPP system have different mapping images of the operating system. Usually, the user submits a job to an administration system that assigns the job to each processor. However, the MPP system allows the user to log onto a specified node, or assign the job to one specified node. Because each node in the MPP system has its own memory that is addressed independently, the data access and communication between the nodes are realized via software techniques.

6.1.4 Beowulf PC Cluster

Beowulf PC clusters are scalable performance clusters based on commodity hardware and a private system network, and utilize open source software (Linux) infrastructure [9]. The commodity hardware can be any number of mass-market and stand-alone computer nodes, even as simple as two networked computers each running Linux and sharing a file system, or as complex as 1,024 nodes with a highspeed, low-latency network.

The Beowulf project first began at the Center of Excellence in Space Data and Information Sciences (CESDIS) in mid-1994 with the assembly of a 16-node cluster developed for the Earth and Space Sciences project (ESS) at the Goddard

Space Flight Center (GSFC). The project quickly spread to other NASA sites, to other research and development labs, and then to universities throughout the world. The scope of the project and the number of Beowulf installations have grown over the years and continue to grow at an increasing rate. The design goal of these types of systems, which utilize commodity hardware such as Intel Pentium4 PC's connected with high-speed networks, is to achieve supercomputer performance at a cost much lower than traditional supercomputer hardware. Some high performance network devices such as Myrinet and Giga.net can further improve the efficiency of the Beowulf PC cluster. The typical architecture of Beowulf PC-cluster is shown in Figure 6.4.

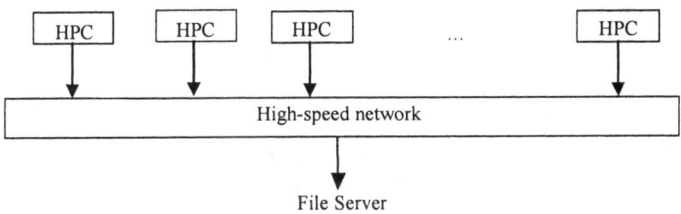

Figure 6.4 Architecture of the Beowulf PC-cluster.

6.2 PARALLEL PROGRAMMING TECHNIQUES

There exist two types of memory addressing techniques in the parallel computing system: the first approach addresses all of the system memory globally, and the second one addresses the local memory individually. According to the memory accessing approach, the parallel programming techniques can be divided into three schemes: message passing such as MPI and parallel virtual machine (PVM), shared memory like OpenMp and pthreads, and the combination of the above two methods. Unlike the message passing scheme, the shared memory scheme is relatively simple because the programmer does not need to consider the location of the data in the memory, progress management, or the barrier operation. The code efficiency is usually not high because it belongs to the fine-grain parallel technique [10] that parallelizes the loop level or statements. In addition, the shared memory parallel code can only run on the shared memory parallel computing system. In contrast to the shared memory parallel technique, although the message passing parallel technique is developed based on the shared memory system, it is more flexible and its code can run specifically on almost all of the parallel computing systems. In the message passing parallel technique, the programmer must take care of the progress management, message passing, and barrier operation. Since the programmer can get more involved in the parallel code design, for example, job assignment, parallel algorithm, and code optimization, this technique has a higher efficiency than the shared memory technique. This is the main reason why there is more software available for the message passing parallel technique than that for the shared memory technique. The implementation of the MPI and PVM is available on most of the popular Beowulf PC clusters, regardless of whether they use 1 GHz Ethernet, Myrinet, or Giga.net. However, we cannot implement the OpenMP in the Beowulf PC cluster because it requires at least 4/8 processors included in the SMP system together with the corresponding programming environment.

6.3 MPICH ARCHITECTURE

At the organizational meeting of the MPI Forum at the Supercomputing '92 Conference, Gropp and Lusk of Mississippi State University volunteered to develop an immediate implementation that would track the standard definition as it evolved. The purpose was to quickly expose problems that the specification might pose for the implementers, and to provide early experimenters with an opportunity to experiment with ideas being proposed for the MPI before they became fixed. The first version of the MPICH implemented the prespecification within a few days after the above meeting. The speed with which this version was completed was due to the existing portable systems P4 and Chameleon. This was the first version of MPICH, which offered a quite reasonable performance as well as portability [11, 12]. The "CH" in the MPICH stands for "chameleon," a symbol of adaptability to one's environment and thus of portability. Chameleons are fast, and from the beginning a secondary goal was to give up as little efficiency as possible for the portability.

The advent of the MPICH has induced significant progress in the implementation of the MPI library. It is based on two facts: (1) the MPICH development is contemporary with the MPI library, and the MPICH implements all of the functions in the MPI library at the same time; and (2) the MPICH has an excellent structure that makes it suitable for any hardware and software environments, as well as useful for the development of other MPI implementation techniques. We will next discuss the MPICH architecture.

The MPICH includes three layers in its architecture. The upper layer is the application programmer interface (API), which carries out the point-to-point communication as well as the collective communication based on the point-to-point communication. The middle layer is the abstract device interface (ADI) in which the device can be simply understood as a bottom-layer communication library. The ADI is a standard that implements different interfaces for the bottom-layer communication library. The bottom-layer is the concrete communication library, for example, P4 and Sockets.

Although the MPI is a huge standard library, it is only remotely related to the hardware device. Using the ADI to implement the MPI, we can provide many shared codes for implementations. Its high performance is realized through the abstract layer provided by the manufacturers. Though the ADI is designed to realize portability of the MPI library, its use is not limited to the MPI library. The ADI in the MPICH can be used to pass a message in the upper layer. The send and receive messages in the ADI are used to pass data between the ADI and hardware, manage the suspend message, and provide some basic information for the execution environment. Particularly, the ADI layer also provides sealing message, adding header information of message, managing the multiple buffers, handling singular message, and matching message. The ADI position in the MPI is shown in Figure 6.5.

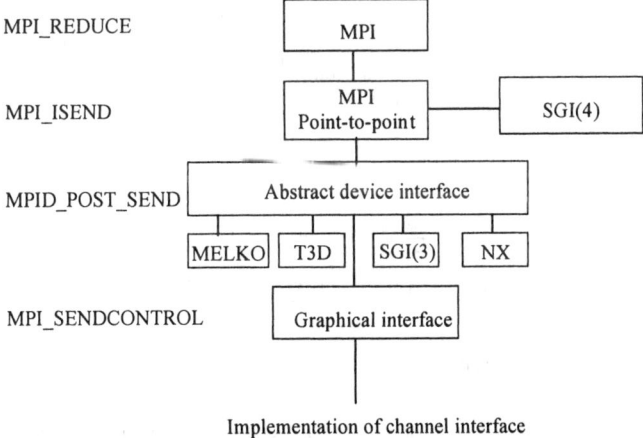

Figure 6.5 ADI in the MPI architecture.

From Figure 6.5 we observe that the ADI provides a control message function similar to the MPID_SendControl. We can realize this function by using the message passing system provided by the manufacturers or through a new code (channel interface).

As the bottom layer of the MPI library, the channel interface only needs to consider how to pass data from one process address to another using the transferring times as little as possible. The channel interface implements five functions: 3 send/receive envelopes (control) including the MPID_SendControl, MPID_RecvControl, and MPID_ControlMSGAvail, and 2 send/receive functions including the MPID_SendChannel and MPID_RecvFromChannel. The MPID-_SendControl in the blocking mode is efficiently realized via the MPID-_SendControlBlock. Similarly, the MPID_SendChannel and MPID_RecvFrom-Channel are also realized through the blocking or nonblocking modes. These functions are implemented on the Unix system via the select, read, and write functions. We utilize the interface buffer technique, and employ the ADI to execute the buffer and control stream management that are usually included in most popular computer systems. Therefore, any additional implementation of buffer management in other layers may degrade the system performance. The structure of the channel interface is shown in Figure 6.6.

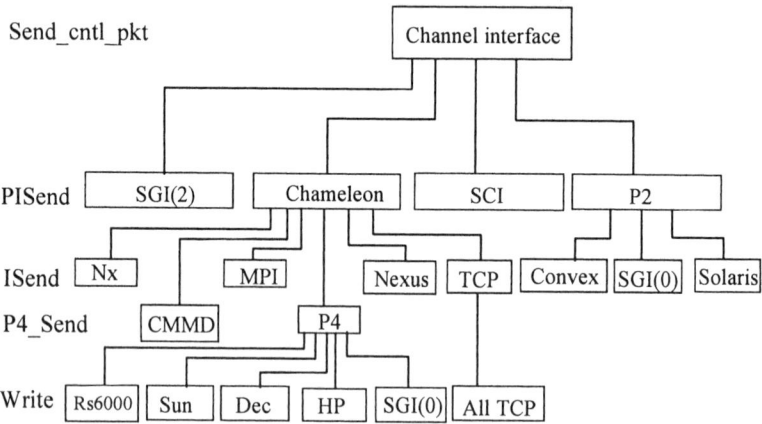

Figure 6.6 Structure of the channel interface.

The channel interface supports three message passing mechanisms. The first, eager, sends a message to a receiver immediately. If the receiver cannot receive this message temporarily, it will create a buffer to store this message. This approach has the best performance and is the default choice of the MPICH. However, if the data in a buffer is too large it may use up all the memory resource. The second, rendezvous, requires the communication between the sender and receiver before sending a message. This is the most reliable approach but it requires the extra time for the communication, hence, its performance is not efficient. The third, get, is similar to rendezvous, however, it requires that the system provide a communication method to realize the message passing from one processor to another. This method relies on the particular hardware such as the shared or nonlocal memory accessing system but it has both the best performance and reliability.

Next, we briefly introduce the definition and classification of the MPI communications. There are four types of communication modes in the MPI: the standard mode, the buffer mode, the synchronous mode, and the ready mode. The main features of communication modes are summarized in Table 6.1.

Table 6.1
Communication Modes

Mode	Completion Condition
Synchronous Send	Only completes when the receive has completed.
Buffered Send	Always completes (unless an error occurs), irrespective of receiver.
Standard Send	Message sent (receiver state unknown).
Ready Send	Always completes (unless an error occurs), irrespective of whether the receive has completed.
Receive	Completes when a message has arrived.

A regular communication mode includes the block and nonblock methods, and its communication operations have the local and nonlocal operations. The definitions of these basic concepts are shown in Table 6.2.

Table 6.2
Definition of Some Concepts

Completion	Memory locations used in the message transfer can be safely accessed.
Block	Return from routine implies completion.
Nonblock	Routine returns immediately and user must test completion.
Local operation	Communication only requires using MPI process instead of communicating with other processes.
Nonlocal operation	Communication needs to communicate with other processes.

6.4 PARALLEL CODE STRUCTURE

As an international standard library, the MPI routine can be executed on different platforms because it deals with processes instead of processors. We need to investigate more details about the concept of process before we discuss the parallel code structure. Existing operating systems today can execute multiple application codes (or processes) at the same time. A regular process can be represented by using four elements, P, C, D, and S, which are the process code, the process control status, the process data, and the process execution status, respectively. When an operating system launches a program, it will create an independent execution environment for the generated process including the memory and address space of commands, program counter, register, stack space, file system, and I/O device. The operating system also assigns the resource to each process according to the specific management algorithm. A process has two basic characteristics: (1) a resource characteristic, which is assigned by the operating system; and (2) an execution characteristic, which indicates the dynamic change of process status during the execution. If a code is executed twice, two independent processes will be created by the operating system since they have different resource characteristics. The five process statuses are shown in Table 6.3.

Table 6.3
Process Status

Status	*Description*
New created status	Process is being created.
Execute status	Process is being executed.
Blocking status	Process is waiting for an event.
Ready status	Process has been called in and is waiting for CPU execute.
Completion status	Process has been completed and system will release the assigned resource for this process.

A process is the minimum unit that the operating system handles. One process has its own memory space, hence, it cannot directly access the memory space allocated for the other processes. There exist three communication modes for the message passing among processes, namely, the communication mode, the synchronization mode, and the gather mode. In the communication mode, a message is passed among processes. If these processes are located in the same processor, the data exchange among the processes can be carried out through the shared data region provided by the system. Otherwise, the message passing among the processes is carried out via network protocols. In the synchronization mode, all the processes wait for each other to keep the same progress, which requires that all the processes satisfy the specific condition before they continue. In the gather mode, the data collected from multiple processes is put together to generate a new data, for instance, summation, minimal, and maximal operations. The new data can be stored as either one process variable or multiple process variables.

Because an operating system assigns the independent resource to each process, the process management is more costly from the computer resource point of view. For example, when an operating system switches from one process to another, it has to back up the resource status of the original process. However, the thread corresponding to the concept of process has the same resource characteristics as the process, but all of the threads share the resource obtained from the system. Therefore, the use of thread can dramatically reduce the cost of the system resource management.

The MPI supports two parallel programming modes: the single program multiple data (SPMD) and the multiple program multiple data (MPMD). A SPMD only includes one executable file that can be executed using the following command:

mpirun –np N <executable file name>

The above command will launch N independent processes that launch the same code on different processors. However, each process executes different commands according to the ID number that the MPI assigns to it. The processes are called into the MPI system via the MPI_Init routine and they pass messages between processes via the MPI routines. There exist two types of communication approaches: namely, point-to-point and broadcast. The processes in the SPMD mode realize the self-identification and communication via a communicator.

Unlike the SPMD mode, the MPMD mode launches multiple executable files using the following command:

mpirun –np N_1 <executable file name>, ..., –np N_n <executable file name>

The total processes are equal to $N = N_1 + ... + N_n$. Although the MPICH library has as many as 200 functions, all of the communications can, in principle, be realized via the use of six basic functions, MPI_Init, MPI_Finalize, MPI_Comm_Rank, MPI_Comm_Size, MPI_Send, and MPI_Recv. The MPI library can be used to develop a variety of parallel computing application codes [13–17], and the main features of parallel processing codes are summarized in Figure 6.7.

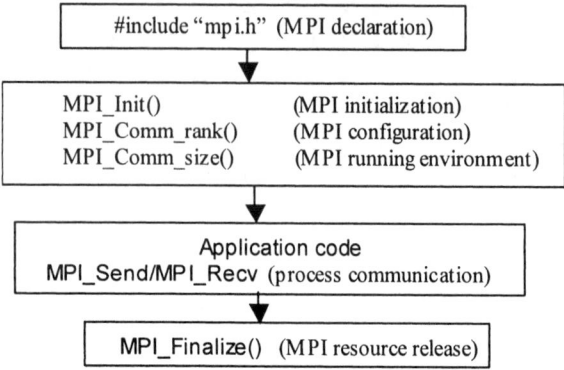

Figure 6.7 Parallel program structure.

Next, we demonstrate the process of parallel code development using the Fortran or C programming language. The area under a sinusoidal function curve (see Figure 6.8) can be expressed as:

$$A = \int_0^\pi \sin(x)\,dx \qquad (6.1)$$

We can assign this job to N processors, and then the area of the shadowed region from x_n to x_{n+1} can be expressed as:

$$A_n = \int_{x_n}^{x_{n+1}} \sin(x)\,dx \approx \sum_{i=n}^{n+1} \sin(x_i)\Delta x_i \qquad (6.2)$$

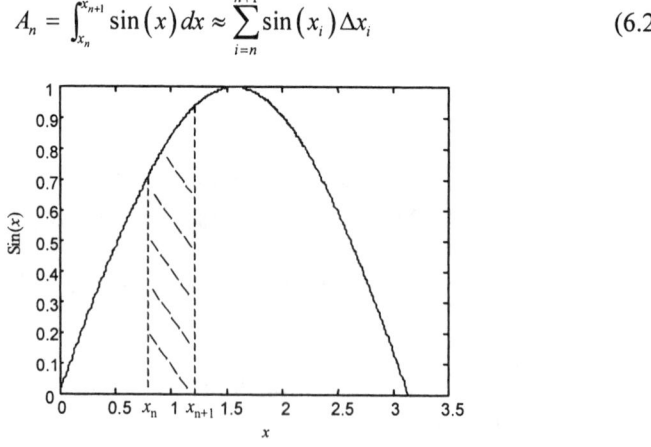

Figure 6.8 Integral of sinusoidal function.

The parallel codes that perform the above calculation in Fortran and C programming languages are listed below. The number of processes will be specified by the user during the code execution. This simple code segment includes the major elements of the parallel processing using the MPI library.

```fortran
Fortran code
program mpi_integral_Fortran
implicit none
! include mpi header file for Fortran
! make sure select fpp option for preprocessor
#include "mpif.h"
real, PARAMETER :: pi = 3.1415926
! Variables
real Integral , sum
integer Index, Size
integer step , ierr, startI, endI, I
real Dx , x
step = 1000
Dx = PI / step
! initialize MPI execution environment
call MPI_Init( ierr )
! get index of the current processor
call MPI_Comm_Rank( MPI_COMM_WORLD, Index, ierr )
! get number of processors
call MPI_Comm_size( MPI_COMM_WORLD, Size, ierr )
! get interval boundary
startI = Index * step / Size;
endI = start + step / Size;
if (endI .gt. 1000 ) endI = 1000;
! calculating
do I = startI , endI - 1
x = Dx * I
integral = Integral + Dx * sin( x )
end do
! add the result from all the processors
call MPI_REDUCE( integral, sum, 1 , MPI_REAL, MPI_SUM, 0 ,
       &                                   MPI_COMM_WORLD, ierr )
! output result
if (Index .eq. 0 ) then
print *, " Result = ", sum
endif
! clean MPI environment
call MPI_Finalize( ierr )
```

end program mpi_integral_Fortran

C code
```c
#include <math.h>
#include <stdio.h>
// include mpi head file
#include <mpi.h>
const float PI = 3.141592654f ;
int main( int argc, char *argv[ ] ) {
float Integral = 0 , sum ;
 int Index, Size ;
 int step = 1000 ;
float Dx = PI / float(step) , x ;
// Initialize MPI execution environment
 MPI_Init( &argc, &argv ) ;
// get index of current processor
 MPI_Comm_rank( MPI_COMM_WORLD, &Index ) ;
// get number of processor
 MPI_Comm_size( MPI_COMM_WORLD, &Size ) ;
// get interval boundary
 int start, end;
 start = Index * step / Size;
 end = start + step / Size;
 if (end > 1000 ) end = 1000;
// calculating
 for( int i = start ; i < end ; i ++ ) {
 x = Dx * i ;
 Integral += Dx * sin( x ) ; }
// add the result from all the processors
 MPI_Reduce( &Integral, &sum , 1, MPI_FLOAT, MPI_SUM, 0
                                  ,MPI_COMM_WORLD) ;
// output result
 if( Index == 0 )
 printf( "Result = %f\n", sum ) ;
// clean MPI environment
 MPI_Finalize( ) ;
 return 0 ; }
```

The code segments above demonstrate a general structure and are not optimized. Hence, they are not suitable for a reference to check the efficiency of parallel codes. In order to test them, it is not necessary to have N processors because the MPI can assign multiple processes to a single processor. In order to run the above codes, the reader needs to set up the proper compiling and running environments including the #include files and library path (see Appendix A for details). For instance, in the Windows platform, assigning the area calculation of A_n to processor N, and specifying the number of processes N in the MPIRun window in Figure 6.9, we click on the Run button and the result will be displayed in the output window. The analytical solution for this example is 2.

Figure 6.9 MPIrun window in the Windows system.

6.5 EFFICIENCY ANALYSIS OF PARALLEL FDTD

In this section, we investigate the efficiency of the parallel FDTD algorithm and the performance of the data exchange on different platforms. The efficiency of a parallel FDTD algorithm depends not only on how we develop the code but also on the computer hardware and networking equipment. We employ the conventional definition of scalability of the parallel processing code [13], which is written as:

$$S = \frac{T_1}{T_n} \qquad (6.3)$$

where T_1 is the simulation time when the entire problem is simulated using a single processor and T_n is the longest simulation time among all processes on n-processors. The efficiency, E, is defined as:

$$E = \frac{S}{n} \qquad (6.4)$$

The first experiment is carried out on the Pennsylvania State University cluster "Lion-xm" [18] (see Figure 6.10). The information of this cluster is given here:

Number of nodes: 128; Number of processors: 256;
Computer: Dell PowerEdge 1750;
Processor: Dual Intel P4 Xeon 3.2 GHz, 1 MB advanced Transfer Cache
RAM: ECC DDR SDRAM (2x2 GB);
NIC: Dual embedded Broadcom 10/100/1000 NICs; Switch: Myricom;

Operating System: Redhat Linux AS 2.1
Fortran Compiler: Intel Fortran 7.0; C++ compiler: Intel C++;
MPI: MPIGM

Figure 6.10 Penn State Lion-xm cluster.

To evaluate the scalability and efficiency defined in (6.3) and (6.4), respectively, we set up an example whose domain size is 300×300×300 cells, and includes a single excitation source and an observation point. In this example, each process is assigned to a physical processor for a fair comparison, even though this is not required by the MPI. The efficiencies obtained in different cases are summarized in Figures 6.11 and 6.12.

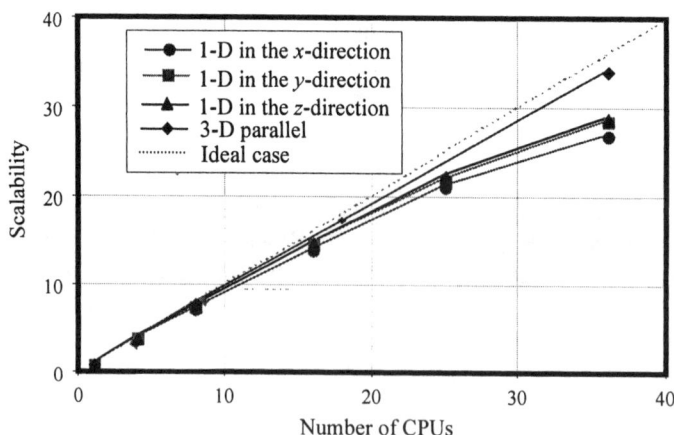

Figure 6.11 Scalability of parallel processing for different processor layout.

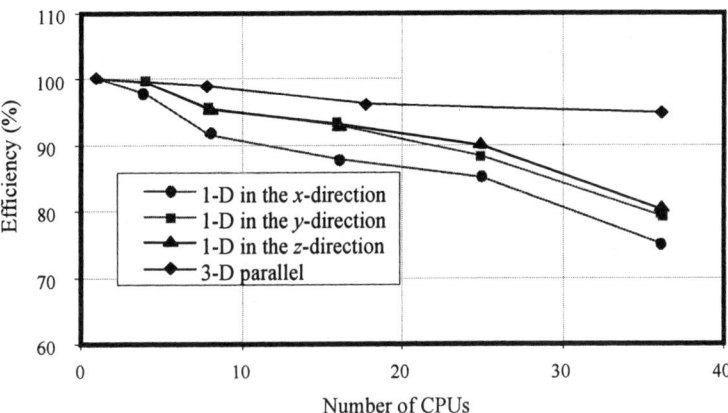

Figure 6.12 Efficiency of parallel processing for different processor layout.

In the FDTD simulations, to get a balanced job distribution we use the PEC boundary to truncate all six walls of the computational domain. The point source and field output at a point do not increase the burden of the parallel FDTD code or affect the efficiency of FDTD code. As a reference, we need to simulate the same job in a single processor to calculate the time T_1 in (6.3). However, this problem is too large to be simulated on a single processor without using virtual memory. Thus, we approximate the time consumption for a single processor by summing the simulation time on four processors, assuming the efficiency in this case is close to 100 percent. The numerical experiments demonstrate that the z-direction division of the subdomains is always better than the x- and y-directions. From Figures 6.11 and 6.12, we observe that the scalability and efficiency of 3-D parallel FDTD code are much higher than that for the 1-D case especially for a large number of processors. The reason for this is that the 3-D domain decomposition contains less interface area size compared to the 1-D cases.

Next, we employ a patch antenna problem to test the efficiency of the parallel FDTD code on a shared memory system [14]. The computational domain including the patch antenna is discretized into $417 \times 833 \times 67$ nonuniform cells. The ground plane is simulated by a PEC boundary and the other five walls are truncated by a six-layer PML. The information of the SGI workstation is given here:

Number of processors: 16
System: SGI Irix 6.5.25f SMP
Processor type: Origin 3400, 16p 400MHz IP35 CPU;
Memory: 16 GB main memory (shared memory)
Compiler: MPI, C++, F90

We use 14 out of 16 processors to simulate the patch antenna problem, and the domain is divided along the direction that has the maximum number of cells. The scalability and efficiency of the parallel FDTD code on the SGI workstation are shown in Figures 6.12 and 6.13.

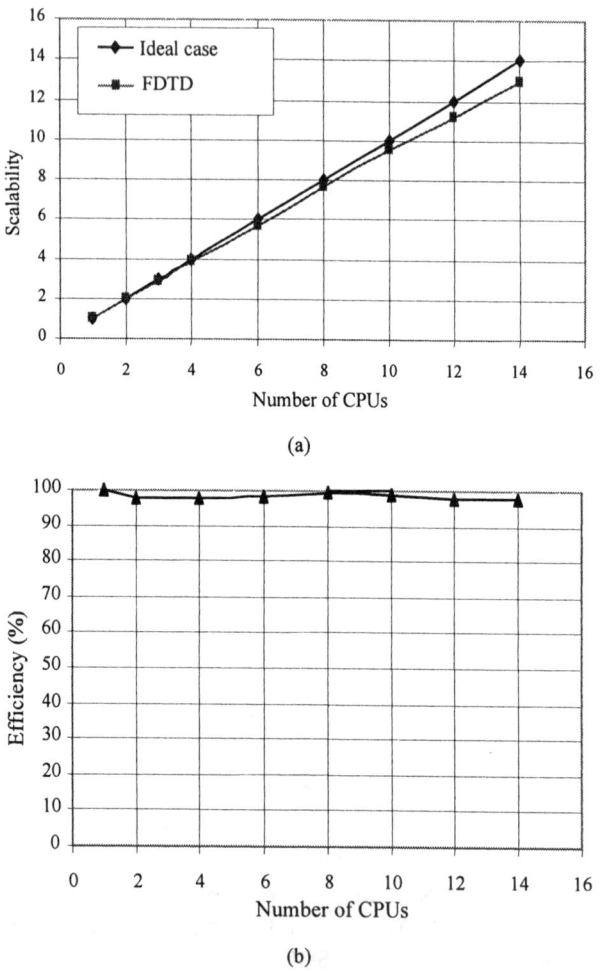

Figure 6.13 Efficiency of parallel FDTD code on SGI workstation. (a) Scalability of parallel FDTD code on SGI workstation. (b) Efficiency of parallel FDTD code on SGI workstation.

Finally, we investigate the efficiency of the parallel FDTD code on the BlueGene/L supercomputer [19, 20], as shown in Figure 6.14. At the time of this writing, this is the fastest computer in the world. The BlueGene/L supercomputer can be used to simulate electrically large problems using up to 65,536 processors via a total 32 TB of memory. The architecture of the BlueGene/L supercomputer is shown in Figure 6.15. The machine that is used for the parallel FDTD test is the LOFAR BlueGene/L located in the Computing Centre of the Rijksuniversiteit Groningen [21].

Figure 6.14 BlueGene/L supercomputer.

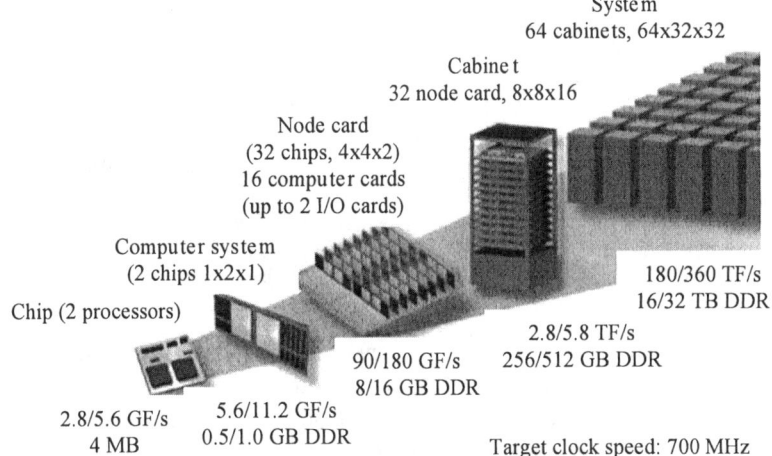

Figure 6.15 BlueGene/L architecture.

Some typical technical parameters of the BlueGene/L supercomputer are shown in Tables 6.4 and 6.5 [19, 20].

Table 6.4
Platform Characteristics of BlueGene/L Supercomputer

Platform characteristics	512-node prototype	64 rack BlueGene/L
Machine peak performance	1.0 / 2.0 TFlops/s	180 / 360 TFlops/s
Total memory size	128 GByte	16 / 32 TByte
Footprint	9 sq feet	2,500 sq feet
Total power	9 KW	1.5 MW
Compute nodes	512 dual proc	65,536 dual proc
Clock frequency	500 MHz	700 MHz
Networks	Torus, Tree, Barrier	Torus, Tree, Barrier
Torus bandwidth	3 B/cycle	3 B/cycle

Table 6.5
Key Features of BlueGene/L Supercomputer

Software Component	Key Feature
Compute node kernel	Scalability via simplicity and determinism
Linux	More complete range of OS services
Compilers: XLF, XLC/C++	Industry standard; automatic support for SIMD FPU
MPI library	Based on MPICH2, highly tuned to BG/L
Control system	Novel, database-centric design

For the problem under test, the efficiency of the parallel FDTD code was found to be 90% with 4,000 processors [22–24]. Specifically, the performance of the code on the 4,000 processors in the LOFAR BlueGene/L is found to be 3,552 times faster than a single processor. Its scalability and efficiency are shown in Figures 6.16(a) and 6.16(b), respectively.

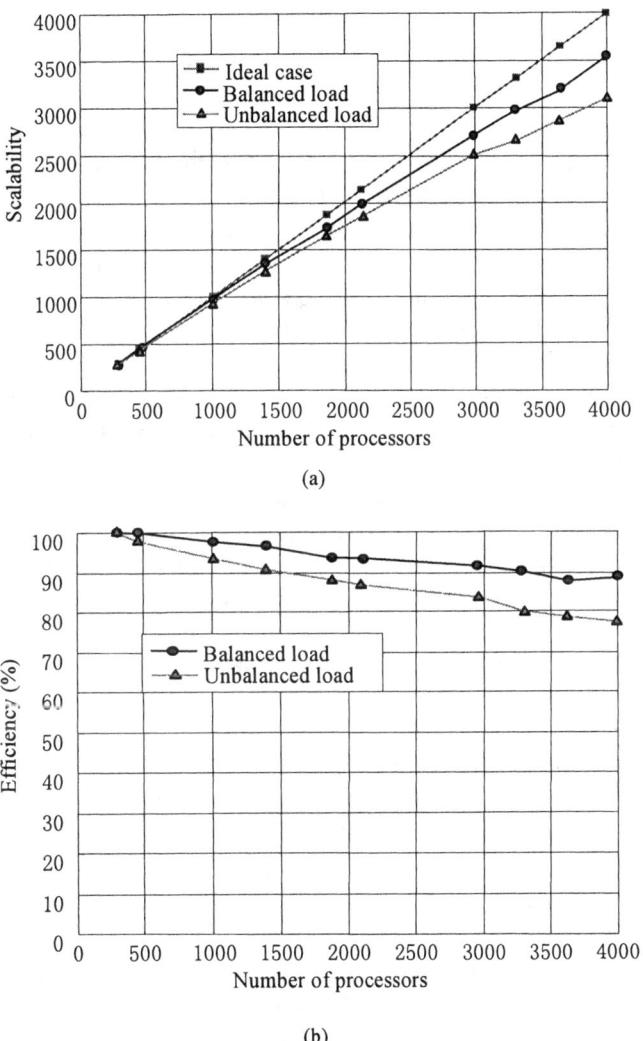

Figure 6.16 Scalability and efficiency of the FDTD code on the LOFAR BlueGene/L. (a) Scalability of the parallel FDTD algorithm. (b) Efficiency of the parallel FDTD algorithm.

The problem is a 100×100 patch antenna array whose dimensions are $1{,}278 \times 1{,}351 \times 43$ mm that are discretized into $1{,}534 \times 1{,}837 \times 74$ nonuniform cells. All six walls of the computational domain are truncated by six-layer PML. We employ up to 4,000 processors to simulate this antenna array problem. It is evident from Figure 6.16 that the efficiency of the parallel FDTD code on the LOFAR BlueGene/L is relatively high compared to the conventional PC clusters.

The tested processor layouts are 288(12×12×2), 450(15×15×2), 800(20×20×2), 1,080(18×28×2), 1,408(22×32×2), 1,872(26×36×2), 2,128(28×38×2), 2,992(34×44×2), 3,312(36×46×2), 3,648(38×48×2), and 4,000(40×50×2), respectively. The simulation results have no relationship with the processor layouts and/or interface locations. Though we can place the interface in the z-direction such that they coincide with the PEC patch, the exchanged data amount can be reduced only in those processors whose interface contains the patch; if one of processors does not include the patch, we do not really get any benefits from this layout. In contrast, if the PML region is assigned to the processors that it touches, it introduces serious a load balancing problem as the number of processors is increased. The curve labeled with "unbalanced load" in Figure 6.16 corresponds to the cases in which the computational domain is truncated by the PML. We note that the efficiency of the parallel FDTD code decreases with the number of processors. In contrast, the curve labeled with "balanced load" in Figure 6.16, in which the computational domain is truncated by the Mur boundary condition, exhibits a higher efficiency.

The outputs generated by the FDTD simulation in this problem include voltage and current at the input port, field distribution on a surface above the patch array in the frequency domain, as well as 2-D and 3-D far-field patterns. If the field distributions are outputted in the text format, the time consumed on the result collection and output becomes comparable to that needed for the FDTD updating. For instance, the FDTD simulation involving 30,000 time steps takes only 38 minutes but the result collection and output (900 Mbytes) takes 11.5 minutes using 4,000 processors. If we use the binary format (120 Mbytes) to output the same result, the time consumed on the result output will reduce from 700 to 44 seconds. It is worthwhile mentioning that the MPICH2 supports the parallel output format, hence, we do not have to collect all the results to the master processor in the MPICH2, but instead write the results from each processor to the hard disk directly. Four commonly used parallel MPI functions in the MPICH2 are MPI_File_write_at (block mode), MPI_File_iwrite_at (nonblock mode), MPI_File_write (block mode), and MPI_File_iwrite (nonblock mode). Both the MPI_File_write_at and MPI_File_iwrite_at have six arguments, which are handle of file, data offset in the output file, data buffer, length of data buffer, data type, and MPI status (output parameter). The functions MPI_File_write and MPI_File_iwrite are similar to MPI_File_write_at and MPI_File_iwrite_at except that they do not have the data offset parameter; however, they require calling another MPI function MPI_File_set_view first in order to set the data offset in the output file.

REFERENCES

[1] M. Flynn, "Some Computer Organizations and Their Effectiveness," *IEEE Transactions on Computers*, Vol. C-21, 1972.

[2] W. Gropp and E. Lusk, *A Test Implementation of the MPI Draft Message-Passing Standard*, Technical Report ANL-92/47, Argonne National Laboratory, December 1992.

[3] M. Snir, et al., *MPI: The Complete Reference*, Vols. I and II, MIT Press, Cambridge, MA, 1998.

[4] W. Gropp, E. Lusk, and A. Skjellum, "A High-Performance, Portable Implementation of the MPI Message Passing Interface Standard," http://www-unix.mcs.anl.gov/mpi/mpich/papers /mpicharticle /paper.html#Node0.

[5] http://publib.boulder.ibm.com/infocenter/pseries/v5r3/index.jsp?topic=/com.ibm.aix.doc/aixbm an/prftungd/multiprocess1.htm.

[6] http://computing-dictionary.thefreedictionary.com/SMP.

[7] D. Walker, *Standards for Message Passing in a Distributed Memory Environment*, Technical Report, Oak Ridge National Laboratory, August 1992.

[8] http://computing-dictionary.thefreedictionary.com/MPP.

[9] D. Becker, et al., (NASA), Beowulf Project: http://www.beowulf.org/.

[10] A. Smyk and M. Tudruj, *RDMA Control Support for Fine-Grain Parallel Computations*, PDP, La Coruna, Spain, 2004.

[11] MPICH (MPI Chameleon), Argonne National Laboratory: http://www.mcs.anl.gov/.

[12] http://www.mpi-forum.org/docs/mpi-20-html/mpi2-report.html.

[13] C. Guiffaut and K. Mahdjoubi, "A Parallel FDTD Algorithm Using the MPI Library," *IEEE Antennas and Propagation Magazine*, Vol. 43, No. 2, April 2001, pp. 94-103.

[14] W. Yu, et al., "A Robust Parallelized Conformal Finite Difference Time Domain Field Solver Package Using the MPI Library," *IEEE Antennas and Propagation Magazine*, Vol. 47, No. 3, 2005, pp. 39-59.

[15] J. Dongarra, I. Foster, and G. Fox, *The Sourcebook of Parallel Computing*, Morgan Kaufmann, San Francisco, CA, 2002.

[16] A. Grama, et al., *Introduction to Parallel Computing*, Addison Wesley, Reading, MA, 2003.

[17] B. Wilkinson and M. Allen, *Parallel Programming: Techniques and Applications Using Networked Workstations and Parallel Computers*, 1st ed., Prentice Hall, Upper Saddle River, NJ, 1998.

[18] http://gears.aset.psu.edu/hpc/systems/lionxm/.

[19] T. Takken, "BlueGene/L Power, Packaging and Cooling," http://www.physik.uni-regensburg.de /studium/uebungen/scomp/BGWS_03_PowerPackagingCooling.pdf.

[20] BlueGene/L Team, "An Overview of the BlueGene/L Supercomputer," *Supercomputing*, November 2002.

[21] http://www.lofar.org/BlueGene/index.htm.

[22] R. Mittra, et al., "Interconnect Modeling on the IBM Blue Gene/L Using the Parallelized Maxwell Solver PFDTD," *IBM Research 6th Annual Austin CAS Conference*, Austin, TX, Feburary 24–25, 2005.

[23] R. Maaskant, et al., "Exploring Vivaldi Element Focal Plane Array Computations Using a Parallel FDTD Algorithm on LOFAR BlueGene/L," *IEEE APS*, Albuquerque, NM, 2006.

[24] W. Yu, et al., "New Direction in Computational Electromagnetics: Solving Large Problem Using the Parallel FDTD on the BlueGene/L Supercomputer Yielding TeraHop-Level Performance," *IEEE Antennas and Propagation Magazine*, 2007 (to appear).

Chapter 7

Parallel FDTD Method

In this chapter, we will introduce the concepts of the parallel FDTD method, which is the principal theme of this book. Generally speaking, parallel computing implies parallel processing, both in the time and spatial domains. In the time domain, parallel processing is implemented via the use of the pipelining technique, in which a task is performed in stages, and the output of one stage serves as the input to the next. This technique speeds up the algorithm by enabling several parts of different tasks to be run concurrently. If the functional units are set up appropriately in the pipeline, then they can yield partial results during each instruction cycle, and this is a typical technique used in most of the existing processors today. In contrast to the time domain parallelism, the spatial parallel technique utilizes a group of processors in a cluster that performs a single task simultaneously. In this book, we will concentrate on only the latter type of parallel processing technique, because it is best suited for parallelizing the FDTD algorithm.

If a given algorithm requires massive data exchange between the nodes of a cluster, then the interconnecting device between the computation elements becomes one of the major bottlenecks that lower the system efficiency, even when the supercomputer is otherwise fast. The FDTD algorithm is inherently parallel in nature because it only requires the exchange of the tangential field components on domain boundaries that are the nearest neighbors, and consequently the parallel FDTD enjoys a very high efficiency. This implies that the parallel FDTD is faster than a serial counterpart almost by a factor N, where N is the number of processors.

Earlier, before the availability of the message passing interface (MPI), most parallel codes were tailored for specific machines. However, the MPI has now become an international standard, and the users have enjoyed its portability and high performance. An important feature of the MPI, which makes it even more attractive, is that it can be downloaded freely from the official MPI Web site. The

MPI library has found extensive applications to a variety of scientific research and solution of engineering problems [1–10]. Following the parallel computing system presented in Chapter 6 in the first section of this chapter, we first briefly introduce the history and features of the MPI library, then focus on the implementation of the parallel FDTD algorithm in the rest of this chapter.

7.1 INTRODUCTION TO THE MPI LIBRARY

As mentioned earlier, the MPI is one of the most widely used parallel processing standards, and it is a library description rather than a new programming language, and its functions can be called from a C, C++, or Fortran program. Prior to the development of the MPI between the 1980s and early 1990s, developers of parallel codes had to choose between portability, efficiency, price, and functionality. The situation improved dramatically, however, when in April 1992, a meeting was held at the American Parallel Research Center, where the basic contents of the MPI library were approved. Next, in a meeting of the MPI working group held in Minneapolis in November 1992, the Oak Ridge National Laboratory submitted a report on the parallel MPI for the very first time. The above conference also created an MPI forum that included 175 scientists and engineers from forty research institutes and universities. The first version of the MPI library was introduced in a supercomputer conference in 1993 and the freely downloadable versions of this library became available from the Web site in May 1994.

The MPI library has the following features:

Standardization	The MPI is the message passing interface library that may be regarded as an international standard. It supports virtually all high-performance computing platforms.
Portability	It is not necessary to modify the source code when an application is ported to different platforms that support the MPI.
Performance	Vendor implementations are able to exploit native hardware features to optimize performance.
Functionality	Includes over 200 routines.
Availability	A variety of implementations are available in the public domain to both the vendors and users.
Standardization	The MPI is the only parallel standard that supports all of the computing platforms.

7.2 DATA EXCHANGING TECHNIQUES

In this section, we describe three of the most widely used schemes that are used for the exchange of data between the adjacent subdomains in the FDTD simulations. Unlike the finite element method (FEM) or the method of moments (MoM), both the electric and magnetic field updates in the FDTD simulations require only the field information from the neighboring cells. Consequently, the FDTD algorithm is inherently parallel in nature and can be easily implemented with the MPI library to run on multiple processors. In parallel processing, the MPI routines are employed for exchanging the tangential electric and/or magnetic fields at points located on the interfaces between adjacent subdomains. It is natural to choose the location of the subdomain boundary so that it coincides with the electric field grid, because then it becomes unnecessary to interpolate these fields in order to derive the tangential electric field components needed for exchanging between the domains.

For the sake of simplicity, we assume that the subdomains are partitioned along one direction, say, the x-direction in Figure 7.1. There exist three commonly used schemes for exchanging the tangential fields between the subdomains required in the parallel FDTD. The first approach is to exchange both the tangential electric and magnetic fields, as shown in Figure 7.1.

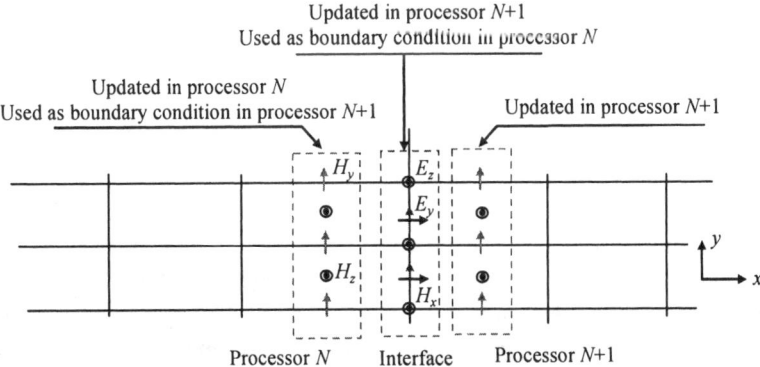

Figure 7.1 Configuration of both the electric and magnetic field exchange.

In this scheme, the tangential magnetic fields H_z and H_y are updated in the subdomain N, and then forwarded to the subdomain $(N+1)$. They are then used as boundary conditions in the subdomain $(N+1)$ to calculate the electric fields E_z and E_y on the subdomain interface. The E_z and E_y fields on the subdomain interface are updated in the subdomain $(N+1)$ and then forwarded to the subdomain N. This procedure, which is repeated at each time step, requires that the electric field be exchanged between two subdomains following the updating of this field. Likewise, the magnetic field is to be exchanged after its updating has been completed, and hence, the field exchange procedure is carried out twice in each time step.

Another alternate and attractive approach is to exchange only the magnetic fields H_z and H_y, which are used to update the electric field E_y on the interface of both the subdomains. These magnetic and electric field locations are depicted in Figure 7.2. Although the electric fields located on the subdomain interfaces are calculated twice, only the magnetic fields at points located adjacent to the interface need to be exchanged at each time step of the FDTD simulation.

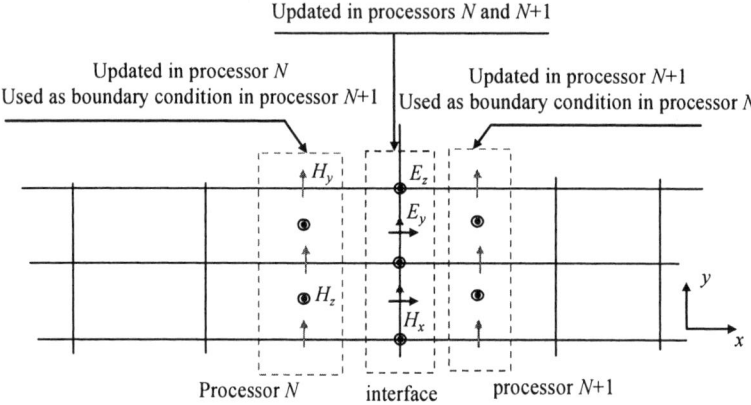

Figure 7.2 Configuration of the magnetic field exchange.

There is a third approach [3, 6] that utilizes a one-cell overlap region to exchange the information between adjacent subdomains, as shown in Figure 7.3. In common with the second scheme, only the tangential magnetic fields are exchanged at each time step. In terms of the parallel FDTD algorithm, all the three schemes are essentially same. However, the third scheme offers a significant advantage over the other two described above for a nonuniform mesh, conformal technique, and inhomogeneous environment. This is because in the third scheme the tangential electric fields on the interface are now located inside the subdomains, instead of on their boundaries. Since the mesh and material information needed for the electric field update are already included in both the subdomains in this scheme, the further exchange of such information is unnecessary during the FDTD iteration. It is also worthwhile mentioning that numerical experiments have shown that the last two schemes are slightly faster than the first one.

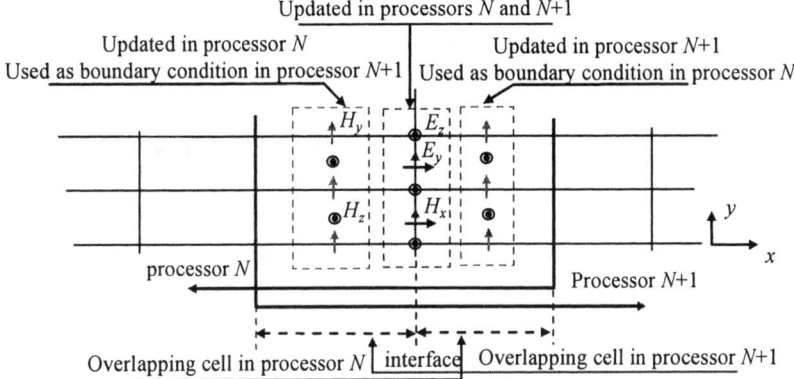

Figure 7.3 Configuration of field exchange with one-cell overlapping region.

In order to develop a robust parallelized FDTD code, we recommend the adaptation of the third scheme and implement the magnetic field exchange between the adjacent subdomains. In this scheme, the subdomain size is two cells larger than its actually physical size for those subdomains located in the middle of the computational domain, while it is one cell larger than its actual size for the first and last subdomains. In common with the second scheme described above, the boundary values being exchanged between the subdomains are the tangential magnetic fields in the third scheme, and are fetched from its near neighbors through the use of MPI routines. In contrast, the electric fields located on the subdomain interface become the inner points, and are updated independently in both the subdomains. Because the electric fields on the subdomain interface are the inner points instead of being located at the boundaries, their updates neither require any mesh information nor the knowledge of material properties from its neighbors. In the FDTD simulations, the magnetic fields are updated concurrently in each subdomain, and the MPI routines are then employed to exchange the boundary values of the magnetic fields. No exchange of electric field information is involved in this parallel processing scheme.

One of the biggest advantages of the overlapping cell is that the mesh generation module becomes much more robust [11] because it becomes unnecessary to generate a conformal mesh on the subdomain boundary (see Figure 7.4(b)). The use of the overlapping region ensures us that the information for the field update in each subdomain will be available within its own region. In addition, it is much more convenient to deal with deformed cells on the subdomain interfaces.

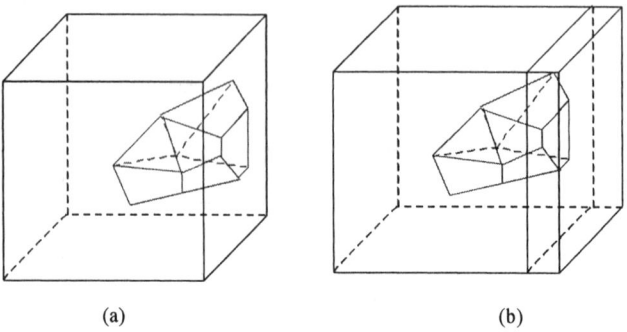

Figure 7.4 Deformed cell location with and without overlapping region. (a) Deformed cell on the subdomain interface without the overlapping cell. (b) Deformed cell on the subdomain interface is inside the subdomain with the overlapping cell.

The data exchange procedure, followed in the third scheme, is illustrated in Figure 7.5. Strictly speaking, it is unnecessary to update the fields indicated in light color to be updated in the FDTD simulations, because they have been updated already in the adjacent neighbor.

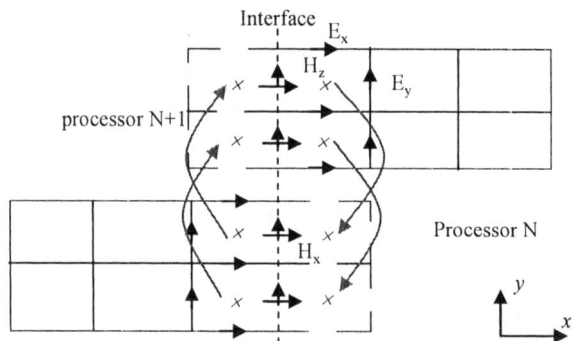

Figure 7.5 Magnetic field communication between adjacent subdomains.

7.3 DOMAIN DECOMPOSITION TECHNIQUE

The 2-D and 3-D parallel FDTD schemes are more flexible than their 1-D counterpart and implementation of these techniques are essentially the same. However, we need to carry out the magnetic field exchange in two or three directions, contemporaneously, in the 2-D and 3-D cases. The parallel processing configurations for the 1-D, 2-D, and 3-D implementations are symbolically shown in Figures 7.6 through 7.8.

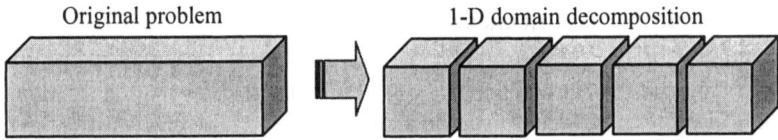

Figure 7.6 1-D parallel processing configuration.

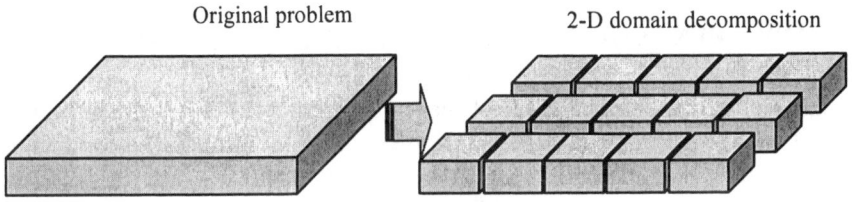

Figure 7.7 2-D parallel processing configuration.

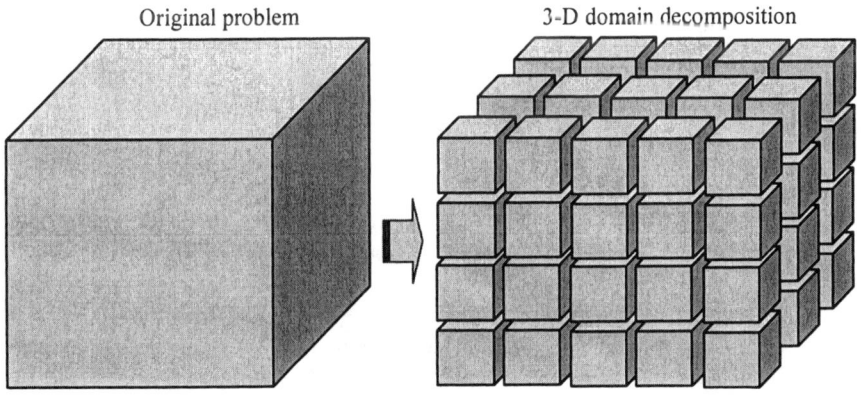

Figure 7.8 3-D parallel processing configuration.

We should clarify the terminology followed in this book, namely, that the nomenclatures 1-D, 2-D, or 3-D parallel FDTD simply imply that the subdomains are partitioned along one, two, or three coordinate directions, respectively, though the FDTD code always deals with three spatial dimensions, namely, x, y, and z.

For a given data exchange scheme, the most important factor that influences the parallel efficiency is the load balancing. The amount of data to be exchanged is directly proportional to the cross-sectional area of the subdomain interface. Hence, the optimal domain decomposition scheme goes hand in hand with the appropriate choice of the processor distribution.

In addition to the field exchange at the subdomain boundary, other FDTD options such as the excitation source, output options, and far-field prediction also will significantly influence the load balancing. However, the optimization of these FDTD options is much more complex than the field exchange because our freedom of choice is much more limited for these options. For instance, we cannot equally assign the current distribution on a surface to individual processors.

7.4 IMPLEMENTATION OF THE PARALLEL FDTD METHOD

In this section, we describe the implementation of the MPI library in the parallel FDTD through a demo code. A demo code segment in Fortran 90 programming language that is used to exchange the magnetic fields between the adjacent subdomains along the *x*-, *y*-, and *z*-directions is shown in the code segment 7.1. It has been validated successfully on various platforms, including Windows, Linux, SGI, UNIX, and the IBM BlueGene/L.

Code segment 7.1
```
      SUBROUTINE PARALLEL_H_EXCHANGE()
      integer i , j , k, tag, count, newType, status(MPI_STATUS_SIZE)
      integer ierr, req(8), tempType, offsets(-1: nz+1)
      integer X_DirCutType, oldType(-1:nz+1), blocklens(-1: nz+1)
      real*4  r1
      real*8  r2
      IF (n_cut_dir_flag .eq. 1) THEN
        IF (single_flag .eq. .true.) THEN
           CALL MPI_Type_vector(ny+3,1,nx+3,MPI_REAL,tempType,ierr)
           CALL MPI_TYPE_COMMIT(tempType, ierr)
           DO i = -1, nz+1
                blocklens(i) = 1
                oldType(i) = tempType
                offsets(i) = (i+1)*sizeof(r1)*(nx+3)*(ny+3)
           END DO
        ELSEIF (double_flag .eq. .true.) THEN
           CALL MPI_Type_vector(ny+3, 1, nx+3, MPI_DOUBLE_PRECISION,
 &                     tempType, ierr)
           CALL MPI_TYPE_COMMIT(tempType, ierr)
```

```
          DO i = -1, nz+1
               blocklens(i) = 1
               oldType(i) = tempType
               offsets(i) = (i+1)*sizeof(r2)*(nx+3)*(ny+3)
          END DO
     END IF
     CALL MPI_TYPE_STRUCT(nz+3, blocklens, offsets, oldType,
    &                         X_DirectionCutType, ierr)
     CALL MPI_TYPE_COMMIT(X_DirectionCutType, ierr)
     IF (ind_process .ne. 0 ) THEN
          CALL MPI_ISend(hy(1,-1,-1),1,X_DirCutType,ind_process-
    &                    1,0,MPI_COMM_WORLD,req(1),ierr)
          CALL MPI_IRECV(hy(0,-1,-1),1,X_DirCutType,ind_process-
    &                    1,1,MPI_COMM_WORLD,req(2),ierr)
          CALL MPI_ISend(hz(1,-1,-1),1,X_DirCutType, ind_process-
    &                    1,2,MPI_COMM_WORLD,req(3), ierr)
          CALL MPI_IRECV(hz(0,-1,-1),1,X_DirCutType,ind_process-
    &                    1,3,MPI_COMM_WORLD,req(4),ierr)
     END IF
     IF (ind_process .ne. num_process-1) THEN
          CALL MPI_IRECV(hy(nx-1,-1,-1),1,X_DirCutType,
    &              ind_process+1, 0,MPI_COMM_WORLD,req(5),ierr)
          CALL MPI_ISend(hy(nx-2,-1,-1),1,X_DirCutType,
    &              ind_process+1, 1,MPI_COMM_WORLD,req(6), ierr)
          CALL MPI_IRECV(hz(nx-1,-1,-1),1,X_DirCutType,
    &              ind_process+1,2,MPI_COMM_WORLD,req(7), ierr)
          CALL MPI_ISend(hz(nx-2,-1,-1),1,X_DirCutType,
    &              ind_process+1,3,MPI_COMM_WORLD,req(8),ierr)
     END IF
     CALL MPI_WAITALL(8, request, status , ierr)
     CALL MPI_TYPE_FREE(X_DirCutType, ierr)
     CALL MPI_TYPE_FREE(tempType, ierr )
     CALL MPI_Barrier(MPI_COMM_WORLD, ierr)
ELSEIF (n_cut_dir_flag .eq. 2) THEN
     IF (single_flag .eq. .true. ) THEN
          CALL MPI_TYPE_VECTOR(nz+3, nx+3, (nx+3)*(ny+3),
    &                         MPI_REAL, newType, ierr)
          CALL MPI_TYPE_COMMIT(newType, ierr)
     ELSEIF (double_flag .eq. .true.) THEN
          CALL MPI_TYPE_VECTOR(nz+3,nx+3,(nx+3)*(ny+3),
    &              MPI_DOUBLE_PRECISION,newType, ierr)
          CALL MPI_TYPE_COMMIT(newType, ierr)
     END IF
     IF (ind_process .ne. 0) THEN
```

```
            CALL MPI_ISend (hx(-1,1,-1),1,newType, ind_process-
    &                       1,0,MPI_COMM_WORLD, req(1), ierr)
            CALL MPI_IRECV(hx(-1,0,-1), 1,newType, ind_process-
    &                       1,1,MPI_COMM_WORLD, req(2), ierr)
            CALL MPI_ISend(hz(-1,1,-1), 1, newType, ind_process-1, 2,
    &                       MPI_COMM_WORLD, req(3), ierr)
            CALL MPI_IRECV(hz(-1,0,-1),1,newType, ind_process-1,3,
    &                       MPI_COMM_WORLD,req(4), ierr)
        END IF
        IF (ind_process .ne. num_process-1) THEN
            CALL MPI_IRECV(hx(-1,ny-1,-1),1,newType,ind_process+1,0,
    &                       MPI_COMM_WORLD,req(5), ierr)
            CALL MPI_ISend(hx(-1,ny-2,-1),1,newType,ind_process+1,
    &                       1, MPI_COMM_WORLD,req(6), ierr)
            CALL MPI_IRECV(hz(-1,ny-1,-1),1,newType,ind_process+1,2,
    &                       MPI_COMM_WORLD,req(7), ierr)
            CALL MPI_ISend (hz(-1,ny-2,-1),1,newType,ind_process+1,3,
    &                       MPI_COMM_WORLD,req(8), ierr)
        END IF
        CALL MPI_WAITALL(8, request, status , ierr)
        CALL MPI_Barrier(MPI_COMM_WORLD, ierr)
        CALL MPI_TYPE_FREE( newType, ierr )
    ELSEIF (n_cut_dir_flag .eq. 3) THEN
    count = ( nx + 3 ) * ( ny + 3 )
    IF (single_flag .eq. .true.) THEN
        IF (ind_process .ne. 0) THEN
            CALL MPI_ISend(hx(-1, -1, 1), count, MPI_REAL,
    &               ind_process-1,0,MPI_COMM_WORLD, req(1), ierr)
            CALL MPI_IRECV(hx(-1, -1, 0), count, MPI_REAL,
    &               ind_process-1, 1, MPI_COMM_WORLD, req(2), ierr)
            CALL MPI_ISend(hy(-1, -1, 1), count, MPI_REAL,
    &               ind_process-1, 2, MPI_COMM_WORLD, req(3), ierr)

            CALL MPI_IRECV(hy(-1, -1, 0), count, MPI_REAL,
    &               ind_process-1, 3, MPI_COMM_WORLD, req(4), ierr)
        END IF
        IF (ind_process .ne. num_process-1) THEN
            CALL MPI_IRECV(hx(-1, -1, nz-1), count, MPI_REAL,
    &               ind_process+1, 0, MPI_COMM_WORLD, req(5), ierr)
            CALL MPI_ISend(hx(-1, -1, nz-2), count, MPI_REAL,
    &               ind_process+1, 1, MPI_COMM_WORLD, req(6), ierr)
            CALL MPI_IRECV(hy(-1, -1, nz-1), count, MPI_REAL,
    &               ind_process+1, 2, MPI_COMM_WORLD, req(7), ierr)
            CALL MPI_ISend(hy(-1, -1, nz-2), count, MPI_REAL,
```

```
     &                        ind_process+1, 3, MPI_COMM_WORLD, req(8), ierr)
                         END IF
                  ELSEIF (double_flag .eq. .true.) THEN
                         IF (ind_process .ne. 0) THEN
                                CALL MPI_ISend(hx(-1, -1, 1), count,
     &                                MPI_DOUBLE_PRECISION, ind_process-1, 0,
     &                                MPI_COMM_WORLD, req(1), ierr)
                                CALL MPI_IRECV(hx(-1, -1, 0), count,
     &                                MPI_DOUBLE_PRECISION, ind_process-1, 1,
     &                                MPI_COMM_WORLD, req(2), ierr)
                                CALL MPI_ISend(hy(-1, -1, 1), count,
     &                                MPI_DOUBLE_PRECISION, ind_process-1, 2,
     &                                MPI_COMM_WORLD, req(3), ierr)
                                CALL MPI_IRECV(hy(-1, -1, 0), count,
     &                                MPI_DOUBLE_PRECISION, ind_process-1, 3,
     &                                MPI_COMM_WORLD, req(4), ierr)
                         END IF
                         IF (ind_process .ne. num_process-1) THEN
                                CALL MPI_IRECV(hx(-1,-1,nz-1),count,
     &                                MPI_DOUBLE_PRECISION, ind_process+1, 0,
     &                                PI_COMM_WORLD, req(5), ierr)
                                CALL MPI_ISend(hx(-1, -1, nz-2), count,
     &                                MPI_DOUBLE_PRECISION, ind_process+1, 1,
     &                                MPI_COMM_WORLD, req(6), ierr)
                                CALL MPI_IRECV (hy(-1,-1,nz-1),count,
     &                                MPI_DOUBLE_PRECISION, ind_process+1, 2,
     &                                MPI_COMM_WORLD, req(7), ierr)
                                CALL MPI_ISend(hy(-1, -1, nz-2), count,
     &                                MPI_DOUBLE_PRECISION, ind_process+1, 3,
     &                                MPI_COMM_WORLD, req(8), ierr)
                         END IF
                  END IF
                  CALL MPI_WAITALL(8, request, status , ierr)
                  CALL MPI_Barrier(MPI_COMM_WORLD, ierr)
           END IF
     END SUBROUTINE PARALLEL_H_EXCHANGE
```

In the demo code above, the local variables nx, ny, and nz are numbers of the FDTD cells along the x-, y-, and z-directions, respectively. The memory allocation range for all of the six field components, ex, ey, ez, hx, hy, and hz, are $(-1:nx+1)$, $(-1:ny+1)$, and $(-1:nz+1)$ along the x-, y-, and z-directions, respectively. The additional indices allocated at both ends are designed to improve the performance of the code. The variables single_flag and double_flag flags are used to control the floating point precision of the field components. The variable n_cut_dir_flag is equal to 1, 2, or 3 for the cases where the subdomains are divided along the *x*-, *y*-, and *z*-directions, respectively. The integer number ind_process is the index of processes and varies from 0 to Num_process-1. In most parallel implementations, each subdomain is assigned to a single processor; hence, the two terminologies, namely "subdomain" and "process," are interchangeably used throughout this book. Furthermore, we often assign each process to a single processor for the most efficient CPU usage, even though this is not required by the MPI. Thus, sometimes the word "processor" is also synonymous with "subdomain."

In the following subsections, we explain the MPI routines, variables, arrays, and data structures used in the code segment 7.1. Although the domain decomposition along the *x*-, *y*-, or *z*-direction is not different from the FDTD point of view, their implementation is very different, owing to different storage structures of the 3-D arrays in the physical memory.

7.4.1 Data Exchange Along the *x*-Direction

Two adjacent subdomains along the *x*-direction are shown in Figure 7.9, where the exchange procedure of the magnetic fields is illustrated.

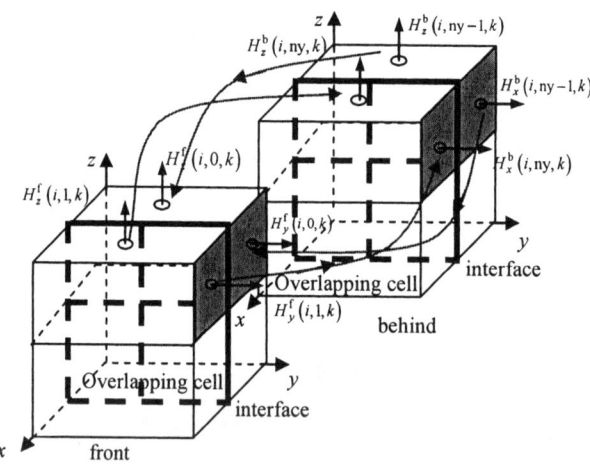

Figure 7.9 Data exchange of two adjacent subdomains in the *x*-direction.

In the Fortran 90 programming language, the first dimension of a 3-D array is continuously stored in the physical memory. Thus the tangential magnetic fields in the *y*-*z* plane are not continuously stored. The total number of field elements in the *y*-*z* plane is (ny+3)×(nz+3) according to the array definition. For the 2-D and 3-D cases, the total element number included in the exchange area depends on the subdomain distribution. Regardless of the type of domain decomposition, two adjacent magnetic field components in the *y*-direction have an index offset of (nx+3) in the physical memory, and this number is the first dimension of the 3-D array. The data structure is illustrated in Figure 7.10, in which the field components to be exchanged are marked as solid circles. This data is used to derive a new data type, called *stride*, for a more efficient exchange procedure.

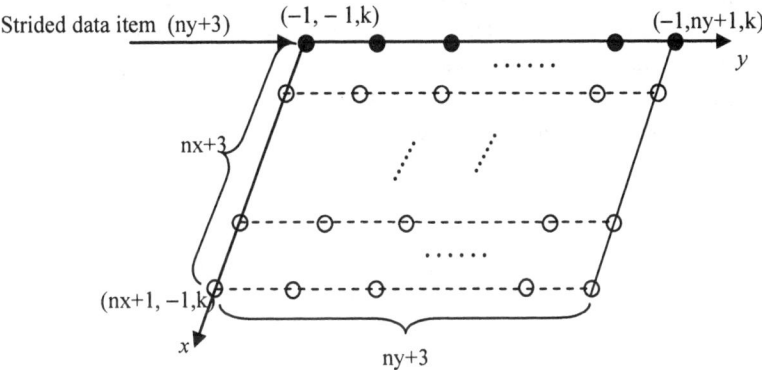

Figure 7.10 Data structure in the *x-y* plane.

We will now explain the MPI routines in the demo code segment 7.1 in the following:

(1) MPI_Type_vector(ny+3, 1, nx+3, MPI_REAL, tempType, ierr)

The MPI routine, MPI_Type_vector, creates a datatype that describes a group of elements separated by a constant amount in memory (a constant stride). The first argument is the number of blocks, which equals (ny+3) in this case and ranges from −1 to ny+1. The second one represents the number of elements in each block. Here, it is simply equal to 1 since each block has only one field element. The third argument is the stride, namely, the separation distance between two blocks. The above three numbers have the same unit that is the datatype specified by the fourth argument. The MPI_REAL is a macro defined for a 4-byte floating point number. The fifth argument is the handler to the derived datatype, an output parameter, which would be exchanged and stored in memory temporarily. The last argument ierr is an error value that returns 0 if the MPI routine is completed successfully.

(2) MPI_TYPE_COMMIT(tempType, ierr)

Any new datatype created by the MPI_Type_vector must be committed to the system with the subroutine MPI_TYPE_COMMIT. This command registers the newly constructed datatype in the system and allows it to fully optimize its operation. Its first argument is a handler of the newly created datatype.

(3) Datatype organization

The first stride is along the *y*-direction and has been created by using the MPI_Type_vector subroutine. The new datatype is tempType and has been used in both the MPI_Type_vector and the MPI_TYPE_COMMIT subroutines. Because the magnetic fields to be exchanged are distributed in the *y-z* plane, and the described datatype above contains only one row of field components, we need to create another stride that includes all the data in the *y-z* plane. In this stride, each block includes one row of data in the *y-z* plane. The offset from one component to another along the *z*-direction is (nx+3)×(ny+3), as shown in Figure 7.11.

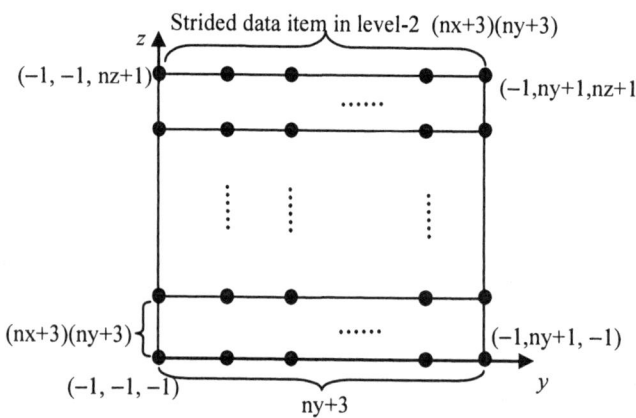

Figure 7.11 3-D data structure deriving a new data type.

Although the original data in the first stride is not successive along the *y*-direction, the newly created datatype, temp_Type, is successive. Therefore, each block in the second stride along the *z*-direction only includes one element of temp_Type. The offest between two successive elements is sizeof(r_1)×(nx+3)×(ny+3), where the variable r_1 is a single precision floating number. The element data type in the second stride is tempType that is created in the first stride.

(4) MPI_TYPE_STRUCT(nz+3, blocklens, offsets, oldType, X_DirectionCut-Type, ierr)

The MPI_TYPE_STRUCT is a data structure and considers the data to be composed of a set of data blocks, each of which has a count and a datatype associated with it, as well as a location given as an offset. The first argument is the number of blocks, which, in this case, is the number of rows of the magnetic field along the z-direction. The second argument is the length of the block. Once again, it just equals 1, because it consists of a single element of the newly created datatype tempType. The third argument is the array of element offset in terms of bytes between two adjacent blocks. The fourth argument is the element datatype that is previously created by the MPI_Type_vector command. The fifth argument, X_DirectionCutType created by the MPI_TYPE_STRUCT, is a datatype of each send buffer element. The last argument ierr is an error value that returns 0 if the MPI routine is completed successfully.

(5) Double precision

The only difference between the single and double precisions is that the element data type of MPI_Type_vector and the offset calculation in the MPI_TYPE_STRUCT are changed from the single to double precision. The macro MPI_DOUBLE_PRECISION stands for an 8-byte floating number.

(6) Sending and receiving a message between processes

As mentioned earlier, the FDTD algorithm is embarrassingly parallel in nature, since both the electric and magnetic field updates in this algorithm only require the information in their adjacent neighbors. The data exchange in the FDTD simulation occurs only on the interfaces between the adjacent subdomains. For the 1-D parallel processing, the first and last subdomains only exchange magnetic fields on the inner boundary, and an inner subdomain needs to exchange magnetic fields with its two near neighbors.

We always use the MPI_ISend and MPI_IRecv to exchange the field data between the subdomains and their neighbors. Compared to the MPI_Send and MPI_Recv, the MPI_ISend and MPI_IRecv are faster because the operation of the MPI_Send and MPI_Recv will not be completed until the message data and envelope are safely stored, when the sender/receiver is free to access and overwrite the send/receive buffer. The messages might be copied either directly into the matching receive buffer or into a temporary system buffer. We can improve the performance on many systems by overlapping the communication and computation. This is especially true on systems where communication can be executed autonomously by an intelligent communication controller. The MPI_ISend and the MPI_IRecv exhibit better performance when they use the nonblock communication, implying that they initiate the send or receive operation, but do not complete them. The operation will be finished before the message has been copied out of the send or receive buffer. A separate send or receive complete call is then needed to complete the communication, namely, to verify that the data has been copied out of the send or receive buffer. Before we introduce the procedure of data exchange from one process to another, let us examine the two MPI routines, MPI_ISend and MPI_IRECV.

(i) MPI_ISend(hy(1, −1, −1), count, X_DirectionCutType, index_processor-1, tag, MPI_COMM_WORLD, request(1), ierr)

The MPI_ISend begins with a nonblock send operation and its arguments are the same as the MPI_Send with an additional handler at the next to the last argument. The two routines behave similarly, except that for the MPI_ISend, the buffer containing the message to be sent cannot be modified until the message has been delivered completely. The MPI_ISend routine has 8 arguments. The first argument is the pointer to the send buffer. The hy(1, −1, −1) component in the outer processor is the second one in the x-direction and it has an index "1" (the first index is 0). Though neither hy(i, −1, −1) nor hy(i,ny+1,nz+1) are used in the update equations, they are still considered as data in the data exchange procedure. The second argument, count, is the number of elements in the send buffer; it equals 1 here because each block includes a successive element in the created datatype. The third argument, X_DirectionCutType created by the MPI_TYPE_STRUCT, is a datatype of each send buffer element. The fourth one, index_processor-1, is the rank of destination. The destination is the inner processor for all of the processors except the first one. For instance, if the indices of the current and previous processors are index_processor and index_processor−1, respectively, the following one will be labeled by the index, index_processor+1. The fifth argument, namely tag, is a matching tag (a nonnegative number) that is used to identify the corresponding relationship between the sender and the receiver. The sixth argument, MPI_COMM_WORLD, is a communicator that combines one context and a group of processes in a single object, which is an argument to data movement operations. The MPI_COMM_WORLD is one of the items defined in the header file "mpi.h." The destination or the source in a send operation refers to the rank of the process in the group identified with the given communicator, MPI_COMM_WORLD. The seventh argument, request(1), is a handler, and it indicates that the send message will be kept in the buffer until the send process has been completed. (Note: The MPI_Send routine does not have this argument.) The last argument, ierr, is an error value. Before the value is returned, the current MPI error handler is called. By default, this error handler aborts the MPI job. The error handler may be changed with MPI_Errhandler_set; the predefined error handler MPI_Errors_Return may be used to cause error values to be returned. If an MPI routine is called successfully, there is no error. In the Fortran programming language, all of the MPI routines are subroutines, and are invoked with the call statement.

(ii) MPI_IRecv(hy(0, −1, −1), count, X_DirectionCutType, index_processor− 1, tag, MPI_COMM_WORLD, request(2), ierr)

The MPI_IRecv has the same number of arguments as the MPI_Isend and even the meaning of the arguments are identical for the two MPI routines. The difference between them, as is evident from their names, is that the MPI_IRecv allows a processor to receive a message from its inner processor whereas the MPI_ISend sends a message instead. The tangential magnetic filed component hy(nx−1, −1, −1) next to the last one in the inner processor, is used as the boundary value that corresponds to the first magnetic field component hy(0, −1, −1) in the current processor. The seventh argument, request(2), is a handler, and indicates that the receive message will be kept in the buffer until the receive process is successfully completed.

(iii) MPI_ISend(hz(1, −1, −1), count, X_DirectionCutType, index_processor − 1, tag, MPI_COMM_WORLD, request(3), ierr)

The MPI_ISend sends the X_DirectionCutType (tangential magnetic field component hz) to the inner processor.

(iv) MPI_IRecv(hz(0, −1, −1), count, X_DirectionCutType, index_processor − 1, tag, MPI_COMM_WORLD, request(4), ierr)

The MPI_IRecv receives the X_DirectionCutType (tangential magnetic field component hz) from the inner processor and uses it as the boundary values in the FDTD simulation.

(7) MPI_TYPE_FREE(X_DirectionCutType, ierr)

The MPI_TYPE_FREE releases the memory allocated for the datatype and X_DirectionCutType.

(8) MPI_Barrier(MPI_COMM_WORLD, ierr)

The MPI_Barrier routine ensures that all of the processes defined in the communicator, namely MPI_COMM_WORLD, have been completed. A barrier is a special collective operation that does not let the process continue until all of the processes in the communicator have called the MPI_Barrier. Its first argument, MPI_COMM_WORLD, is defined within the send and receive routines.

(9) Sending and receiving a message from the outer processor

The procedures for sending a message to as well as receiving from the outer processors are similar to the ones used for the inner processors. However, there are two major differences between the two: first, the last processor needs only to exchange a message with the inner one next to it; second, the fourth argument in both the MPI_IRECV and the MPI_ISend routines changes to be index_processor+1.

7.4.2 Data Exchange Along the *y*-Direction

If the cutting flag, n_cut_dir_flag, equals 2, the processors are assigned in the *y*-direction. Though the data exchange procedures are similar in the *x*-, *y*-, and *z*-directions, the data structure is totally different along different directions because the memory of a 3-D array in Fortran programming language is allocated in the different ways along different directions. The data exchange process along the *y*-direction is illustrated in Figure 7.12.

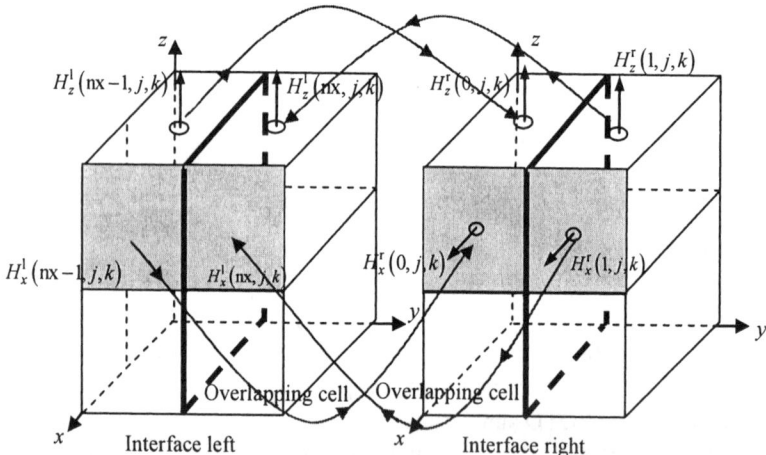

Figure 7.12 Data exchange configuration for processes along the *y*-direction.

The data allocation along the *x*-direction is successive but this is not the case along the *z*-direction in the *x-z* plane. The exchanged data structure and the derived data type are shown in Figure 7.13. The solid points are used to derive a new data type. The data allocation is successive along the *x*-direction and the total number of node is (nx+3)×(ny+3) in the *x-y* plane.

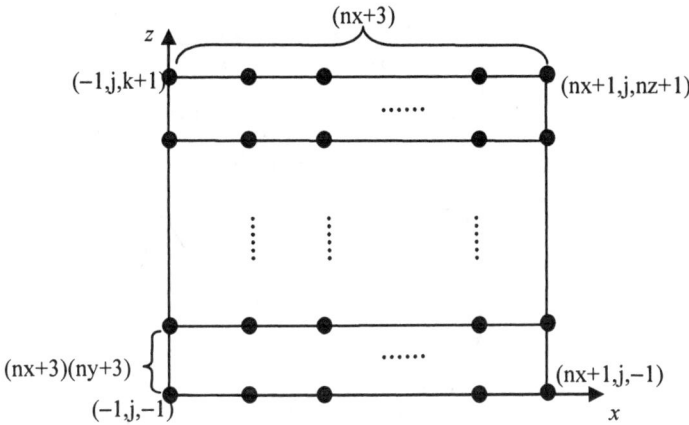

Figure 7.13 3-D data structure for deriving a new data type.

Next, we investigate the MPI subroutine, MPI_TYPE_VECTOR (nz+3, nx+3, (nx+3)×(ny+3), MPI_REAL, newType, ierr). When the subdivision is along the *y*-direction, the data exchange is successive along the *x*-direction, hence, the second argument of the MPI_TYPE_VECTOR, the block length, is (nx+3). The block number, the first argument of MPI_TYPE_VECTOR routine, is (nz+3) because there are (nz+3) groups of data along the *z*-direction. The spacing of the stride, the third argument in the MPI_TYPE_VECTOR, is (nx+3)×(ny+3), which is the distance from the first element of one block to the next one. The rest of the arguments in the MPI_TYPE_VECTOR routine are the same as those used for partitioning in the *x*-direction.

Once we construct the new stride, newType, the message exchanging procedure is the same as that partitioning along the *x*-direction.

7.4.3 Data Exchange Along the *z*-Direction

Compared to the case where the computational domain is subdivided in the *x*- or the *y*-direction, the decomposition along the *z*-direction is the simplest case, because the memory allocation for a 3-D array is successive in the *x-y* plane, and the data in the *x-y* plane can be simply used to construct a data block. For the processors along the *z*-direction, we do not even need to create any new data type because we only have one block in the data structure (see Figures 7.14 and 7.15).

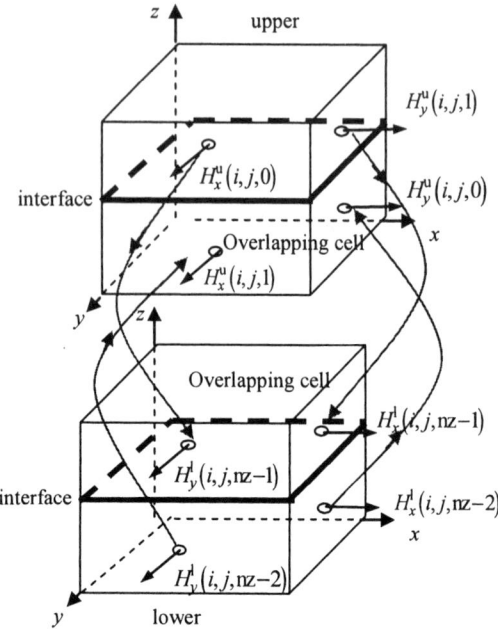

Figure 7.14 Data processing for processes along the z-direction.

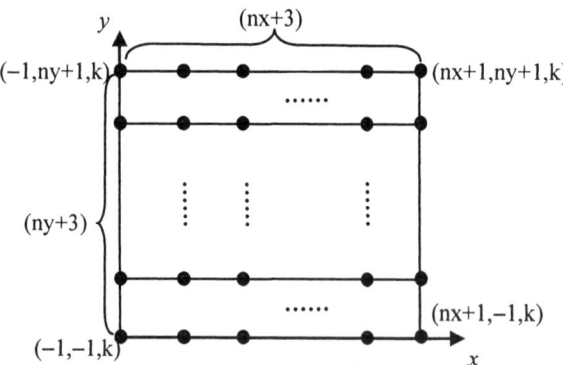

Figure 7.15 Data structure in the x-y plane.

Although the explanation given above is for the 1-D parallel case in which the processors are arranged in one spatial direction, the data exchange scheme can be easily applied to 2-D and 3-D parallel cases. In these parallel cases, we only need to adjust the corresponding indices to identify the neighboring subdomains, as well as the starting and ending indices of the field data to be exchanged.

7.5 RESULT COLLECTION

So far we have introduced the data exchanging procedure for the magnetic and/or electric fields in the parallel FDTD method. In this section, we will discuss some relative topics such as mesh and result collections as well as gathering the far-field information.

7.5.1 Nonuniform Mesh Collection

The mesh generation procedure in parallel processing is not substantially different from that employed in a single processor, as long as the range of index is specified in each subdomain. It is important to recognize that the mesh generation scheme in parallel processing must accommodate the situation that the object spans more than one subdomain, but only meshes the part of the object that is located inside the subdomain. The overlapping region as shown in Figure 7.3 ensures that the cell information needed for the electric field update on the interface is included in both processors, and that the conformal mesh is not located on the subdomain boundary in order to develop a robust mesh code.

In the following example, we assume that the computational domain containing a circular horn is subdivided into eight subdomains. The conformal mesh distributions in the second and fourth subdomains are shown in Figure 7.16(a). The MPI routines are used to collect the mesh information from different processors into a single data file in the master processor. The global mesh distribution of the circular horn is shown in Figure 7.16(b).

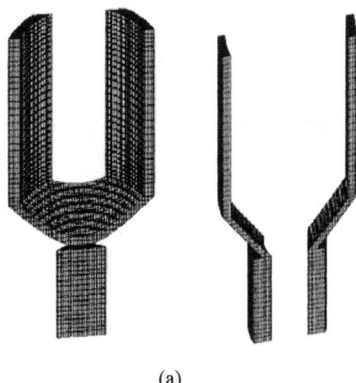

(a)

Figure 7.16 Conformal mesh collection in parallel processing. (a) Individual conformal mesh distribution. (b) Global conformal mesh distribution.

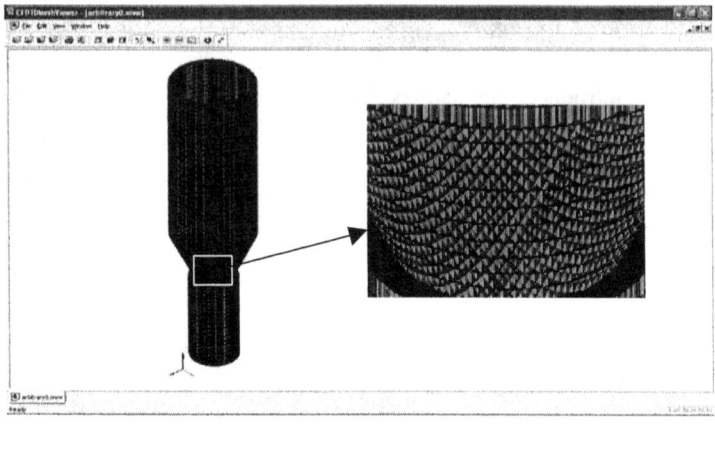

(b)

Figure 7.16 (continued.)

7.5.2 Result Collection

If an observation point is located completely inside one subdomain, then all of the six field components are measured in the corresponding processor. Let us consider the alternative situation where the observation point is located at the interface between two subdomains. The normal electric field component at the lower-bound processor is not available (see Figure 7.5). Therefore, if an observation point is located at the interface between adjacent processors, we should output the field in the upper-bound processor.

If the voltage or current measurement path is located entirely within a single subdomain, then the other subdomains do not contribute to the result of these measurements. However, when a measurement path crosses the interface between the subdomains, the approach to measure the voltage and current is the same as that in the case of a single processor though they are not complete results, and then the MPI_Reduce routine is used to collect the contribution from all of the relevant subdomains. In some special cases, a part of the measurement path resides on the interface, and in this event, the contribution can be measured from both processors, although only one of these should be retained to avoid duplication. A sample code used to collect the voltage into the master processor (ind_process = 0) is shown in the code segment 7.2. The variable voltage is used to store the voltage contribution from each processor, and the voltage_total is the global voltage output. The MPI routine, MPI_ALLREDUCE, is employed to collect the voltage from each processor and place the global voltage in the master processor afterwards.

Code segment 7.2
```
        IF (voltage_flag .eq. "IN_DOMAIN") THEN
            DO k = k_start, k_stop
                        voltage = voltage + dz(k) * ez(i0,j0,k)
            END DO
        ELSE
            voltage = 0.0
        END IF
        CALL MPI_ALLREDUCE(voltage, voltage_total, 1, MPI_REAL,
     &              MPI_SUM, MPI_COMM_WORLD, ierr)
        IF (ind_process .eq. 0) write(voltage_file,*') n, voltage_total
```

7.5.3 Far-Field Collection

In parallel processing, the far-field pattern needs to be computed by aggregating the contributions from all the processors that contain parts of Huygens' box. To derive the equivalent electrical current on the Huygens' surface, we need to know the magnetic fields at both sides of Huygens' surface, as well as the cell size information when a nonuniform mesh is used. If we use overlapping cells, we can circumvent the need for any further data exchange to perform the far-field calculation. However, we need to keep in mind that when a surface of the Huygens' box resides on an interface, its contribution should only be retained from one side and not both.

7.5.4 Surface Current Collection

We will now describe the collection of surface currents and field distributions using the format shown in Figure 7.17 when the specified surface is across the multiple processors. In the parallel FDTD simulation, each processor will produce the surface current distribution in its own region. In order to generate a global surface current distribution, we need to pass the local current or field onto the master processor. Because the 2-D arrays of the local current or field are not stored continuously in the physical memory allocated for the global array, they need to be reorganized in order that they can be stored in the correct order.

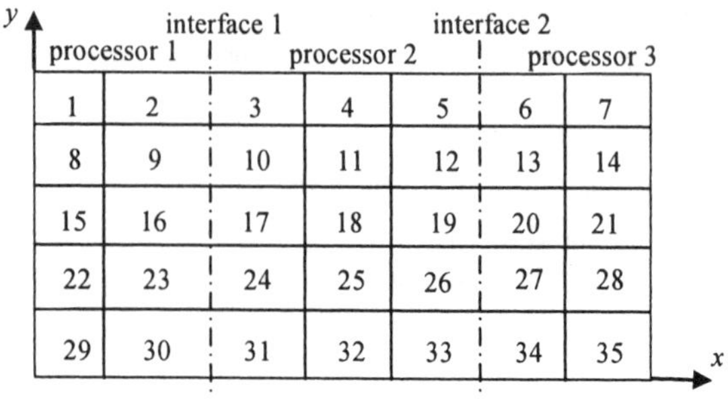

	interface 1			interface 2		
processor 1			processor 2			processor 3
1	2	3	4	5	6	7
8	9	10	11	12	13	14
15	16	17	18	19	20	21
22	23	24	25	26	27	28
29	30	31	32	33	34	35

Figure 7.17 Surface current distribution in the global domain.

The following subroutine (see code segment 7.3) in Fortran 90 is employed to collect the surface current distribution. In this subroutine, the indices, total_nx, total_ny, and total_nz, indicate the total number of cells in the global domain. The nx_dim, ny_dim, and nz_dim are the dimensions of the surface current distribution in each subdomain. The global distribution of the surface electric or magnetic current is collected onto the master processor (ind_process = 0).

Code segment 7.3
```
    SUBROUTINE FIELDINFO_COLLECTION( dim1, dim2 ,d_flag)
    integer disp, ierr, d_flag, i,j,k,dim1,dim2, sendcount(2) , recvcount(2)
    integer, allocatable :: recvcnts(:), displs(:), sendcnt
    real, allocatable:: parallel_temp_for_Xcut(:)
    sendcount(1) = dim1
    sendcount(2) = dim2
    sendcnt = sendcount(1)*sendcount(2)
    allocate(recvcnts(1:num_process), displs(1:num_process))
    IF( n_cut_dir_flag .eq. 1 ) THEN    !cut in the x-direction
        CALL MPI_ALLReduce ( sendcount(1), recvcount(1), 1,
  &         MPI_INTEGER, MPI_SUM, MPI_COMM_WORLD, ierr)
        CALL MPI_Allgather( sendcnt , 1 , MPI_INTEGER , recvcnts, 1,
  &         MPI_INTEGER, MPI_COMM_WORLD,ierr)
        recvcount(2) = sendcount(2)
        allocate ( parallel_temp_for_Xcut(1: recvcount(1)*recvcount(2)))
        DO i = 1, num_process
            displs(i) = 0
            DO j = 1, i -1
                displs(i) = displs(i) + recvcnts(j)
            ENDDO
        ENDDO
        CALL MPI_GATHERV(parallel_temp3,sendcount(1)*sendcount(2),
```

```
&          MPI_REAL, parallel_temp_for_Xcut, recvcnts,displs,MPI_REAL,
&          0, MPI_COMM_WORLD, ierr)
         IF( ind_process .eq. 0 ) THEN
           DO j = 1, recvcount(2)
             DO k = 1, num_process
               IF( k .ne. num_process ) THEN
                 DO i = 1, ( displs(k+1)-displs(k)) / recvcount(2)
                   disp=displs(k)+(j-1)*recvcnts(k)/ recvcount(2)
                   parallel_temp(displs(k)/recvcount(2)+i,j) =
&                         parallel_temp_for_Xcut(i+disp)
                 ENDDO
               ELSE
                 DO i = 1 , recvcnts(k) / recvcount(2)
                   disp = displs(k)+(j-1)*recvcnts(k)/ recvcount(2)
                   parallel_temp(displs(k)/recvcount(2)+i,j) =
&                         parallel_temp_for_Xcut(i+disp)
                 ENDDO
               ENDIF
             ENDDO
           ENDDO
         ENDIF
         deallocate ( parallel_temp_for_Xcut)
       ELSEIF ( n_cut_dir_flag eq. 2 ) THEN    !cut in the y-direction
         CALL MPI_ALLReduce ( sendcount(2), recvcount(2), 1 ,
&             MPI_INTEGER, MPI_SUM, MPI_COMM_WORLD, ierr)
         CALL MPI_Allgather(sendcnt,1,MPI_INTEGER,recvcnts,1,
&             MPI_INTEGER,MPI_COMM_WORLD ,ierr )
         recvcount(1) = sendcount(1)
         DO i = 1 , num_process
           displs(i) = 0
           DO j = 1 , i -1
             displs(i) = displs(i) + recvcnts(j)
           ENDDO
         ENDDO
         CALL MPI_GATHERV(parallel_temp3,sendcount(1)*sendcount(2),
             MPI_REAL, parallel_temp, recvcnts, displs,
&            MPI_REAL, 0, MPI_COMM_WORLD, ierr )
       ELSEIF( n_cut_dir_flag .eq. 3 ) then    !cut in the z-direction
         recvcount(1) = sendcount(1)
         recvcount(2) = sendcount(2)
         CALL MPI_REDUCE(parallel_temp3,parallel_temp,
&            sendcount(1)*sendcount(2), MPI_REAL, MPI_SUM, 0,
&            MPI_COMM_WORLD, ierr)
       ENDIF
```

```
deallocate( recvcnts, displs)
END SUBROUTINE FIELDINFO_COLLECTION
```

As an example, we again consider the horn antenna mentioned previously, which is simulated by using 16 processors with the domain decomposition arranged along the y-direction. The field distribution above the circular horn antenna is calculated separately in all the processors. The global time domain electric field distribution on the surface above the horn antenna is collected using the code segment 7.3 and is displayed in Figure 7.18.

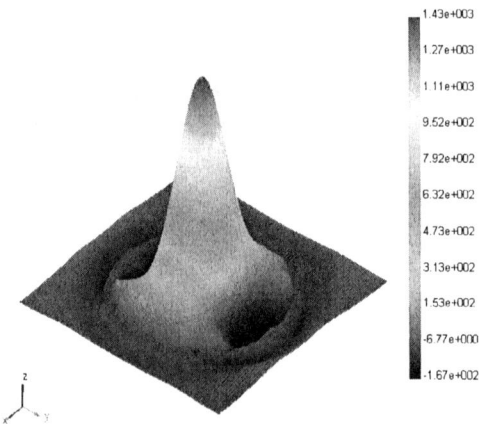

Figure 7.18 Time domain electric field distribution on the surface above the antenna.

7.6 ASSOCIATED PARALLEL TECHNIQUES

In this section, we describe some parallel processing techniques used in the FDTD simulations such as excitation source, boundary condition, waveguide terminator, and subgridding.

7.6.1 Excitation Source

We choose the parallel scheme, in which the tangential electric fields on the subdomain interface are updated at both sides. Therefore, if an excitation source lines up with the subdomain interface, it should be added into two adjacent subdomains. We introduce several frequently used types of excitation sources next.

7.6.1.1 Current Source

Since the electric charge distribution does not explicitly appear in the FDTD update equations, the excitation source in the FDTD simulation will either be the electric or the magnetic current. If an excitation source is completely included in one subdomain, its operation would be very similar to that in the serial FDTD code. However, if an electric current source straddles an interface, or if a magnetic current resides on an interface, the excitation source should be included in both of the two subdomains.

7.6.1.2 Electric Current Source with an Internal Resistance

We now turn to a current source with an internal resistance. The internal resistance is modeled in the FDTD by assigning a finite conductivity to the FDTD cells in the excitation region. The conductivity value is calculated from the length and cross section area of the excitation, and the load resistance. For a matching load with a resistance R that is distributed in an area S and a length of L, we simply calculate the equivalent conductivity distribution σ by using the following formula:

$$R = \frac{L}{\sigma A} \tag{7.1}$$

The area A is taken to be the effective area of the grids, which includes an additional one-half cell in the horizontal direction. For instance, if the resistance R is along the z-direction, the index range of the lumped source is from i_{min} to i_{max} in the x-direction, j_{min} to j_{max} in the y-direction, and k_{min} to k_{max} in the z-direction. Then the area A and length L are defined as:

$$A = \left(\sum_{i=i_{min}}^{i_{max}} \frac{\Delta x(i) + \Delta x(i-1)}{2} \right) \left(\sum_{j=j_{min}}^{j_{max}} \frac{\Delta y(j) + \Delta y(j-1)}{2} \right) \tag{7.2}$$

$$L = \sum_{k=k_{min}}^{k_{max}} dz(k) \tag{7.3}$$

7.6.1.3 Waveguide Excitation

In common with a current source that is introduced in the electric field update, the waveguide excitation also needs to be updated in both subdomains when it resides on the subdomain interface. In a rectangular waveguide excitation, if the width of the waveguide is W_{guide} and $A_0(n)$ is the value of the excitation pulse at the time step n, the excitation source can be expressed as follows:

$$J_y^n(i,j,k) = A_0(n)\sin\left(\frac{x(i)}{W_{guide}}\pi\right) \quad (7.4)$$

In the circular waveguide simulation, the TE_{11} mode has an electric field distribution shown in Figure 7.19. The two electric field components, E_ρ and E_φ, in the polar coordinate system can be expressed as [12]:

$$E_\rho = \frac{1}{\varepsilon\rho} J_1(\beta_\rho \rho)\sin(\varphi) \quad (7.5a)$$

$$E_\varphi = \frac{\beta_\rho}{\varepsilon} J'_1(\beta_\rho \rho)\cos(\varphi) \quad (7.5b)$$

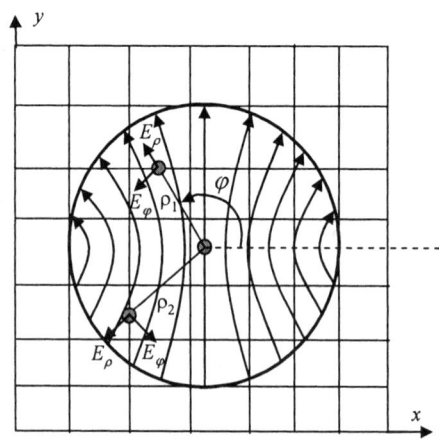

Figure 7.19 Circular waveguide excitation configuration.

Only those electric fields that reside inside the waveguide and on the FDTD grids should be considered in the FDTD simulation. We determine the field values by measuring the distance between the location of the electric field and the central point (reference point). The electric fields in the Cartesian coordinate system, E_x and E_y, are calculated from the following formulas:

$$E_x(i,j,k) = E_\rho \cos(\varphi) - E_\varphi \sin(\varphi) \quad (7.6a)$$

$$E_y(i,j,k) = E_\rho \sin(\varphi) + E_\varphi \cos(\varphi) \quad (7.6b)$$

7.6.1.4 Plane Wave Source

A plane wave source is implemented independently in each subdomain that shares the same global reference point. It is the same as that used in the serial code except that the excitation on the interface should be included in both of the subdomains.

7.6.2 Waveguide Matched Load

Matched loads in waveguides are modeled by using a terminator, which utilizes the first-order Mur's boundary condition [13, 14]. If a terminator is contained either entirely within a single subdomain or straddles an interface, it can be handled as a regular Mur's boundary condition. However, if the boundary condition is to be applied on the interface between two subdomains, it should be implemented only in the subdomain that contains the front side of the terminator. This is because the overlapping scheme does not exchange any electric field information in the overlapping cell, so that only the subdomain that contains the front side of the terminator has the required field data.

7.6.3 Subgridding Technique

If a structure has fine features, a subgridding technique such as [15] can be used to increase the spatial resolution locally. The field exchange between regular grids and subgrids is performed inside an overlapping region, as shown in Figure 7.20. The ADI technique can be used to increase the time step inside the subgridded region so that the FDTD grids and subgrids can share the same time step size.

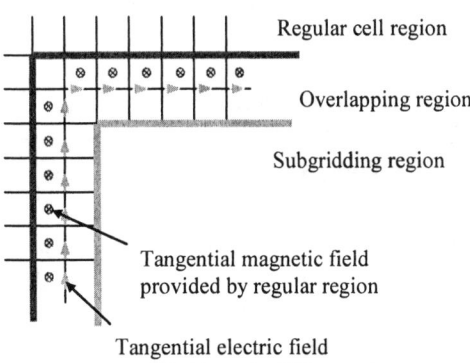

Figure 7.20 Overlapping region between regular FDTD and subgridding regions.

Since the ADI technique is an implicit method and requires the solution of matrix equations, we assign the updating of the fields within the subgridded region to a single processor. The actual job assignment will be much more involved if a problem includes multiple subgridding regions. In this scheme, if a subgridded region resides in more than one subdomain, as shown in Figure 7.21, the field exchange data needs to be collected from and distributed to all the related processors.

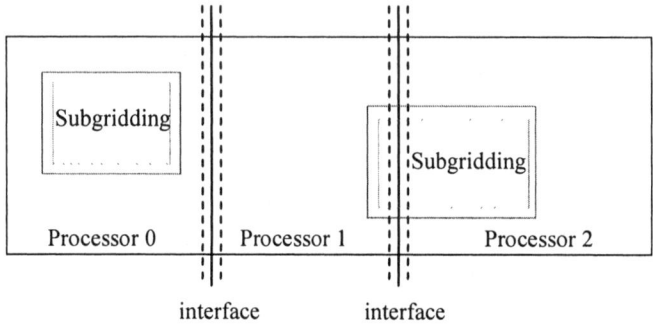

Figure 7.21 Subgridding implementation in the parallel processing.

Before closing this section, we would like to mention the input and output structures in the parallel processing. Each processor has its own input and output project folders. These folders are totally isolated from each other even though they may be located within the same hard disk. Access to the data from one processor to another can only be realized through the MPI routines. In our design, the mesh code is separated from the FDTD code, so the mesh code will create input and output folders for each processor and a common output folder to store the global output results. Each processor will read the project files from its own input folder and place the intermediate results in its own output folder. All of the global results will be stored in the global output folder.

7.7 NUMERICAL EXAMPLES

The efficiency analysis of the parallel FDTD algorithm in different systems has been examined in Chapter 6. In this section, we will validate the parallel processing techniques described in this chapter through three representative examples. These examples are not very large so it is not really necessary to use multiple processors to handle them. However, we use them to simply validate the parallel processing techniques.

7.7.1 Crossed Dipole

The first example we consider is the crossed dipoles shown in the inset of Figure 7.22. The dipoles are excited in phase at the gaps in their centers. The computational domain is discretized into 54 × 42 × 80 uniform cells. The dipole length is 40 mm (one-quarter wavelength at 1 GHz) and the distance between the two dipoles is 75 mm. The output of this problem is the axial ratio of the radiated field, as shown in Figure 7.22. The simulation result shows excellent agreement with those obtained analytically [14].

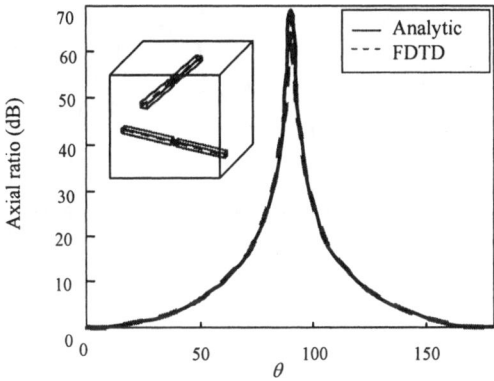

Figure 7.22 Axial ratio of the crossed dipoles.

7.7.2 Circular Horn Antenna

Next, we investigate a circular horn antenna, shown in Figure 7.23(a), which has been designed to achieve low sidelobes. This horn can be used as a feed for reflector antenna systems and is particularly suitable for narrow directional beam forming. The horn is excited in the waveguide region by using a TE_{11} mode. The region between the waveguide and the horn (i.e., the throat) generates the higher-order TM_{11} mode because of the presence of the discontinuity. The angle and length of the transition are chosen to adjust the level of the TE_{11} and TM_{11} mode, while the length of the larger diameter waveguide is designed to control the phase of the modes in order to cancel the electric field from the two modes at the aperture boundary. This in turn helps eliminate the edge current in the conductor that causes sidelobes in the pattern, and produces a nearly circularly symmetric beam. We chose the horn dimensions from the literature [16] whose computational domain is discretized into 87 × 87 × 72 uniform cells. The far-field pattern of the antenna is shown in Figure 7.23(b).

We cannot expect a high efficiency in parallel processing for small problems, especially for an unbalanced partitioning. A more efficient approach is to separate the near-to-far-field transformation module from the FDTD code so that we can avoid the unbalanced job assignment associated with the Huygens' box. To do this, we can store the equivalent surface currents in the time domain on the surface of the Huygens' box into a data file, and then carry out the far-field calculation independently using these stored surface currents.

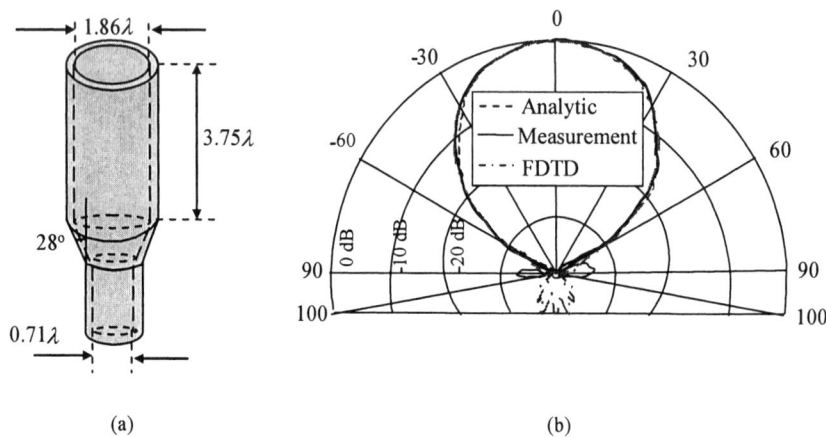

Figure 7.23 Far-field pattern of the circular horn antenna at 12 GHz. (a) Horn configuration. (b) Far-field pattern of horn antenna.

7.7.3 Patch Antenna Array Beam Scanning

For the last example we choose the configuration of a patch antenna array, shown in Figure 7.24. Each element of the array is driven by two coaxial cables to realize circular polarization. The computational domain is discretized into 476 × 466 × 16 nonuniform cells to coincide with the borders of objects. Despite the fact that the problem structure is quite complex, the parallel processing implementations of the FDTD update, mesh generation, excitation source, and result collection can all be handled without difficulty if we follow the techniques described in this chapter.

Since the feed cables are infinite in the vertical direction we cannot place a "closed" Huygens' box to predict the far field. Instead, the far-field pattern is calculated from the aperture field distribution above the patch array, since we are only interested in the far field in the upper hemisphere. In Figure 7.25 we observe slight differences in the large scan angles between the measurement and the simulation results that are attributable to the fact that the Huygens' surface, which we are working with, is finite, whereas, ideally it should be infinite. The 0° and 30° beam scan patterns are shown in Figures 7.25(a) and 7.25(b).

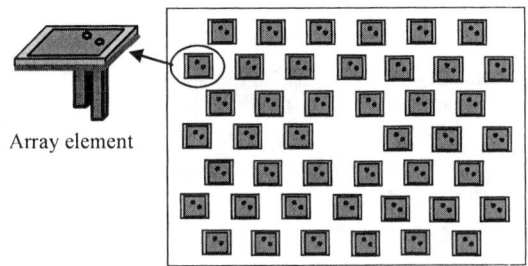

Figure 7.24 Configuration of patch antenna phased array.

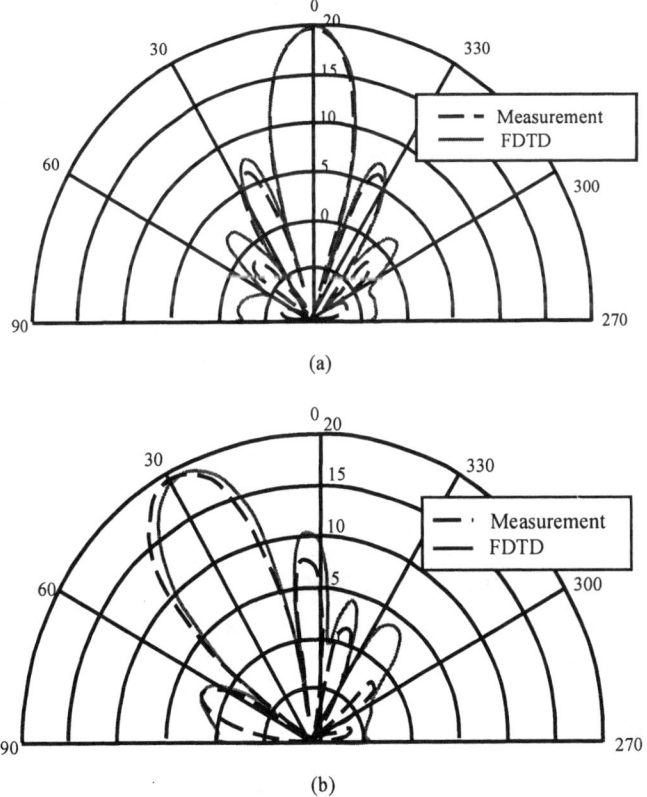

Figure 7.25 (a) 0° and (b) 30° scan radiation patterns of patch array.

REFERENCES

[1] W. Gropp, E. Lusk, and A. Skjellum, *Using MPI: Portable Parallel Programming with the Message-Passing Interface*, 2nd ed., MIT Press, Cambridge, MA, 1999.

[2] MPICH (MPI Chameleon), Argonne National Laboratory: http://www.mcs.anl.gov/.

[3] W. Yu, et al., "A Robust Parallelized Conformal Finite Difference Time Domain Field Solver Package Using the MPI Library," *IEEE Antennas and Propagation Magazine*, Vol. 47, No. 3, 2005, pp. 39-59.

[4] W. Yu, et al., *Parallel Finite Difference Time Domain*, Press of Communication University of China, Beijing, 2005, (in Chinese).

[5] V. Varadarajan and R. Mittra, "Finite-Difference Time-Domain (FDTD) Analysis Using Distributed Computing," *IEEE Microwave and Guided Wave Letters*, Vol. 4, No. 5, May 1994, pp. 144-145.

[6] S. Gedney, "Finite Difference Time Domain Analysis of Microwave Circuit Devices in High Performance Vector/Parallel Computers," *IEEE Transactions on Microwave Theory and Techniques*, Vol. 43, No. 10, October 1995, pp. 2510-2514.

[7] K. Chew and V. Fusco, "A Parallel Implementation of the Finite Difference Time Domain Algorithm," *International Journal of Numerical Modeling Electronic Networks, Devices and Fields*, Vol. 8, 1995, pp. 293-299.

[8] C. Guiffaut and K. Mahdjoubi, "A Parallel FDTD Algorithm Using the MPI Library," *IEEE Antennas and Propagation Magazine*, Vol. 43, No. 2, April 2001, pp. 94-103.

[9] J. Wang, et al., "Computation with a Parallel FDTD System of Human-Body Effect on Electromagnetic Absorption for Portable Telephones," *IEEE Transactions on Microwave Theory and Techniques*, Vol. 52, No. 1, January 2004, pp. 53-58.

[10] L. Catarinucci, P. Palazzari, and L. Tarricone, "A Parallel FDTD Tool for the Solution of Large Dosimetric Problems: An Application to the Interaction Between Humans and Radiobase Antennas," *IEEE Microwave Symposium Digest*, Vol. 3, June 2002, pp. 1755-1758.

[11] T. Su, et al., "A New Conformal Mesh Generating Technique for Conformal Finite-Difference Time-Domain (CFDTD) Method," *IEEE Antennas and Propagation Magazine*, Vol. 46, No. 1, December 2003, pp. 37-49.

[12] C. Balanis, *Antenna Theory*, Harper & Row, New York, 1982.

[13] G. Mur, "Absorbing Boundary Conditions for the Finite-Difference Approximation of the Time-Domain Electromagnetic Field Equations," *IEEE Transactions on Electromagnetic Compatibility*, Vol. 23, No. 3, 1981, pp. 377-382.

[14] W. Yu and R. Mittra, *Conformal Finite-Difference Time-Domain Maxwell's Equations Solver: Software and User's Guide*, Artech House, Norwood, MA, 2004.

[15] M. Marrone, R. Mittra, and W. Yu, "A Novel Approach to Deriving a Stable Hybridized FDTD Algorithm Using the Cell Method," *IEEE International Symposium on Antennas and Propagation*, Columbus, OH, June 2003.

[16] P. Potter, "A New Horn Antenna with Suppressed Sidelobes and Equal Beamwidths," *Microwave Journal*, Vol. 4, June 1963, pp. 71-78.

Chapter 8

Illustrative Engineering Applications

In this chapter we employ the parallel FDTD method to simulate two problems of practical interest, both involving finite arrays. The first example is that of microstrip patch antenna array, while the second one is comprised of crossed dipole elements. Neither of these can be simulated by using a single processor with 2 gigabytes of memory. Though we have chosen array problems to illustrate the application of the parallel FDTD, it can be used to simulate any electrically large problem provided we have access to a sufficient number of processors. Access to a large cluster with a large number of nodes is becoming more and more common these days, and the trend is continuing unabated.

8.1 FINITE PATCH ANTENNA ARRAY

We begin the validation of the parallel FDTD algorithm with the problem of a single patch antenna, shown in Figure 8.1(a). First, we assume that both the dielectric layer and the PEC ground plane are finite in the horizontal directions. The dimensions of the patch and probe are indicated in Figure 8.1(b).

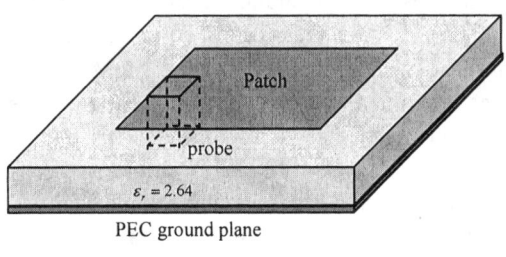

(a)

Figure 8.1 Configuration of single patch antenna. (a) Single patch antenna. (b) Patch structure. (c) Side view of patch antenna.

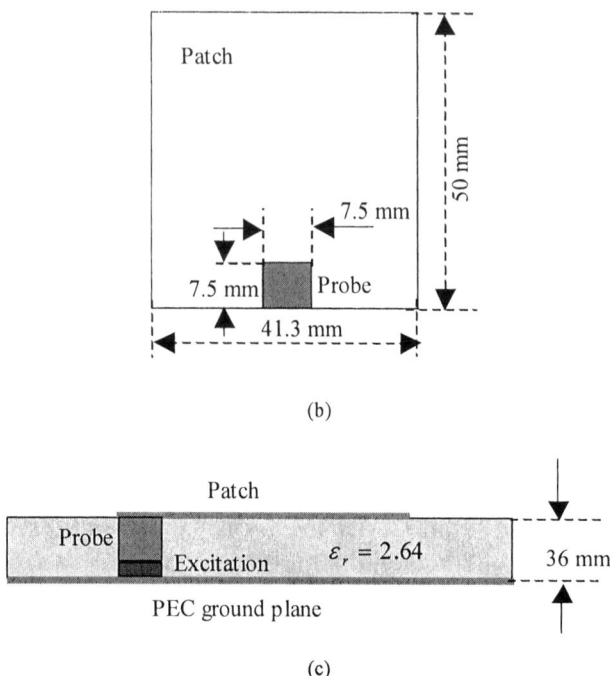

(b)

(c)

Figure 8.1 (continued.)

The Huygens' surface that is used to predict the far-field pattern is placed two cells above the patch antenna and has the same dimensions as the computational domain in the horizontal directions. Even though a single processor is adequate for the FDTD to simulate this problem, we still use the parallel FDTD code and multiple processors to simulate it in order to validate the parallel processing detailed in Chapters 6 and 7. The result simulated by using different computational electromagnetic techniques such as parallel FDTD [1], MoM [2, 3], and FEM [4], are all plotted in Figure 8.2, for the sake of comparison. The infinite dielectric layer and PEC ground plane are used in the MoM simulation because the commercial MoM code we have used was unable to handle either a finite dielectric layer or a PEC ground plane that is also finite. This explains why we see a noticeable difference between the MoM results and those derived using the finite methods that do simulate a finite substrate and ground plane. On the other hand, we observe the excellent agreement between the parallel FDTD and the FEM over the entire frequency range of 1.6 to 2 GHz for which the simulation was carried out.

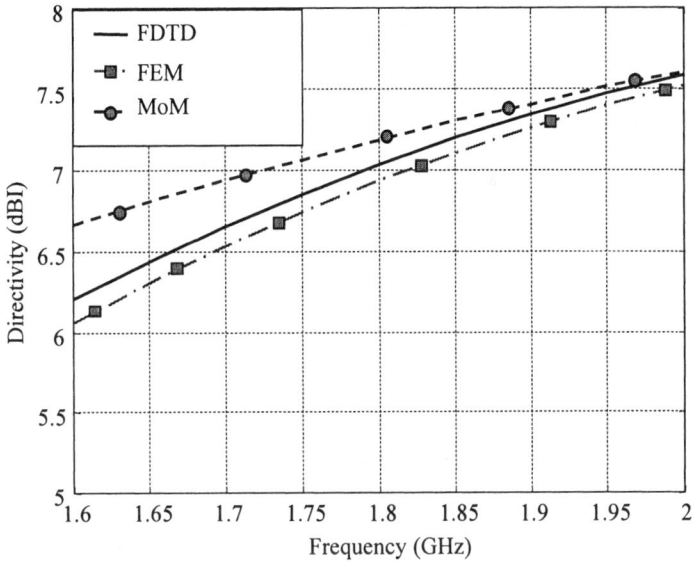

Figure 8.2 Directivity of the single patch antenna.

The array problem that we wish to simulate is comprised of 100 × 100 patch elements. However, neither the FEM nor the MoM is able to solve such a large problem. Therefore, we proceed to simulate a 2 × 2 patch antenna array to demonstrate the capability of the parallel FDTD method before we deal with the full problem. The configuration of the 2 × 2 patch array is shown in Figure 8.3.

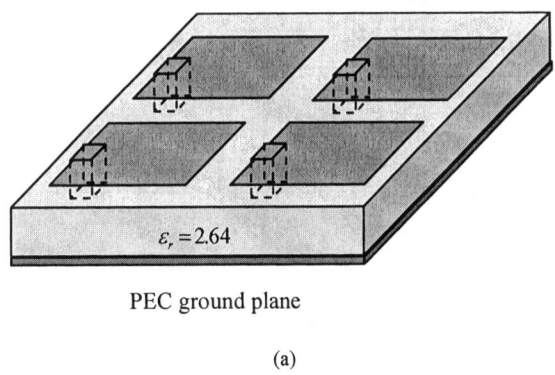

PEC ground plane

(a)

Figure 8.3 Configuration of 2 × 2 patch array. (a) Configuration of 2 × 2 patch array. (b) Dimension of 2 × 2 patch array.

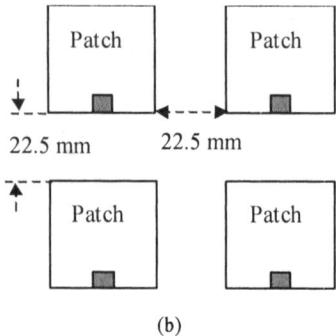

Figure 8.3 (continued.)

The parallel FDTD method is employed to simulate the 2×2 patch array above and the simulation result is plotted in Figure 8.4. Once again, we have also plotted the results simulated by using the FEM and the MoM in the same figure for the sake of comparison. It is worthwhile mentioning that the simulation result is independent of the number of processors if we follow the guidelines prescribed in Chapter 7. The processor distribution is usually determined by the nature of the problem being simulated and for example, a 2-D processor distribution is usually employed for the planar structures.

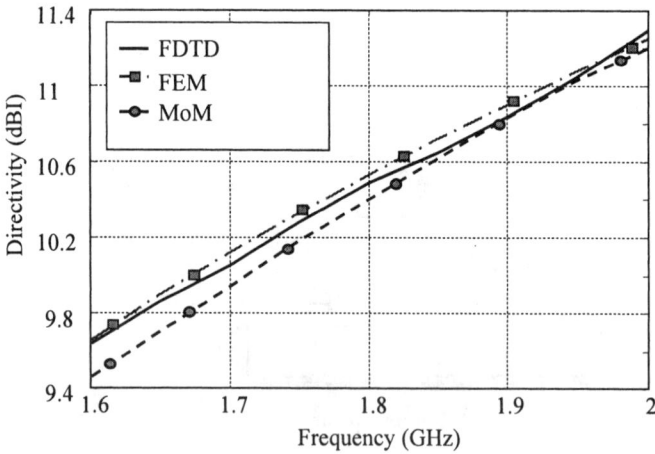

Figure 8.4 Directivity of 2×2 patch array using the different approaches.

The relative locations of the antenna elements in a practical array design significantly affect its performance [5]. The directivity of a 2 × 2 patch array is shown in Figure 8.5 as a function of the separation distance between the elements.

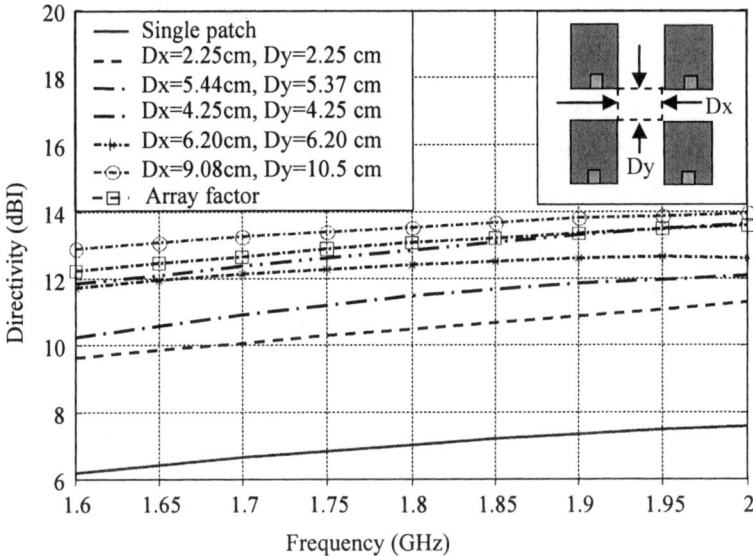

Figure 8.5 Variation of directivity of the 2 × 2 patch array with separation distance.

Next, we return to the full 100 × 100 patch array problem. The dimensions of the patch antenna array are 6,424 mm, 7,286.25 mm, and 3 mm in the x-, y-, and z-directions, respectively. Its electrical size is 48λ at the frequency 2.0 GHz. The computational domain is discretized into 806 × 1,005 × 222 nonuniform cells.

We use 50 processors of the Penn State Lion-xm cluster to simulate this antenna array, and the directivities in $\varphi = 0°$ and $\varphi = 90°$ are shown in Figure 8.6. Once again, the processors can be arranged in 1-D or 2-D distribution, and the simulation results of the parallel FDTD remain independent of our choice of the distribution. As mentioned previously, the problem is too large for the commercial FEM and MoM codes at our disposal; hence, only parallel FDTD results are plotted in the above figure.

There may be as many as 97 sidelobes in the far-field pattern for this large antenna array [5]. Consequently, if there are 8 samples per wavelength the angular resolution of the far field should be at least 0.5°. We use a 0.25° resolution in order to achieve sufficient accuracy in the near-to-far-field transformation, which is rather time consuming for this problem owing to the high angular resolution and the oversampled FDTD mesh on the Huygens' surface. In fact, the time it takes to implement the near-to-far-field transformation is even much longer than that required to carry out the FDTD update.

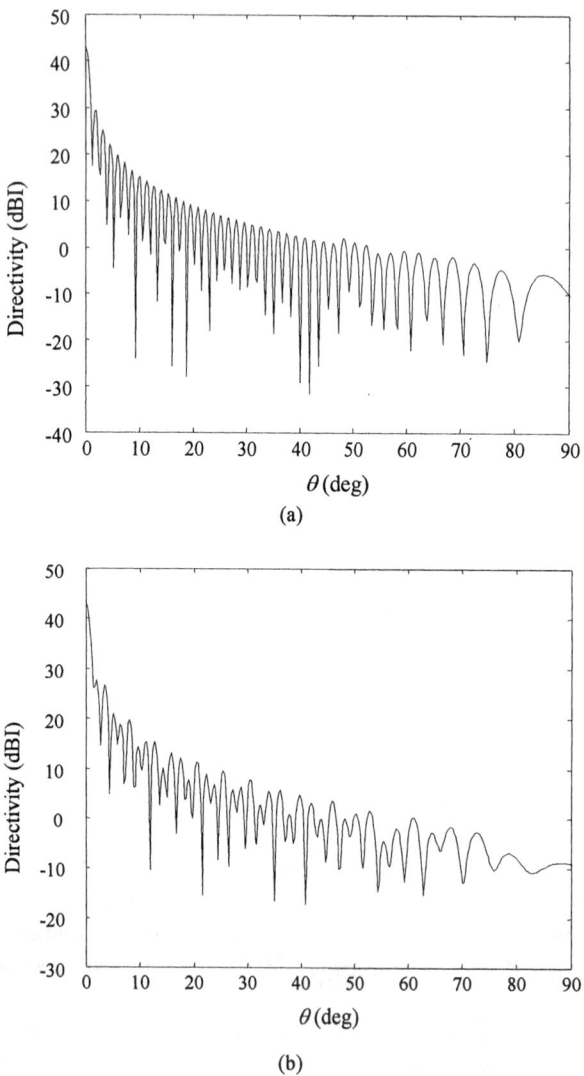

Figure 8.6 Directivity of 100×100 patch array at frequency 1.8 GHz. (a) Directivity at $\varphi = 0°$. (b) Directivity at $\varphi = 90°$.

As mentioned earlier, to calculate the directivity, we need to know the radiated power and the direction of the maximum radiation. The radiation power can be easily computed from integration of the Poynting vector on the Huygens' surface instead of getting it via the far-field calculation, which requires calculating it at each angle and frequency. In order to calculate the maximum radiation direction in the far-field pattern, we first calculate the far field using a relative low angular resolution, and then search for the accurate maximum radiation direction accurately using a high angular resolution over a small range. To this end, we first use a 2° resolution to find the maximum beam-steering direction in the 3-D space, and then use a 0.25° resolution for a more accurate search in the range of 4° around the maximum direction. Using the procedure described above can reduce the time consumed for the directivity calculation to a reasonable amount.

8.2 FINITE CROSSED DIPOLE ANTENNA ARRAY

In this section we use the parallel FDTD method to simulate a crossed dipole array, as shown in Figure 8.7(a). The dipole elements in the array are supported by the thin dielectric slabs that are attached to a large finite PEC ground plane, as shown in Figure 8.7(b). The dipole element is excited by a pair of voltages at the gap of dipole. The antenna elements are organized in a rectangular lattice. Since the width of dipole arms and thickness of the dielectric slabs are very small in terms of the wavelength, the small cell sizes must be employed to achieve accurate results using the FDTD simulation.

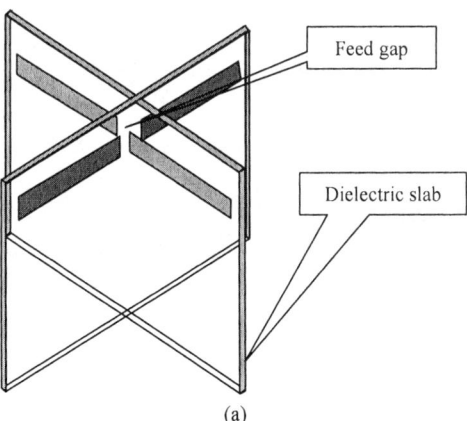

Figure 8.7 Crossed dipole array. (a) Antenna element. (b) 2 × 4 antenna array.

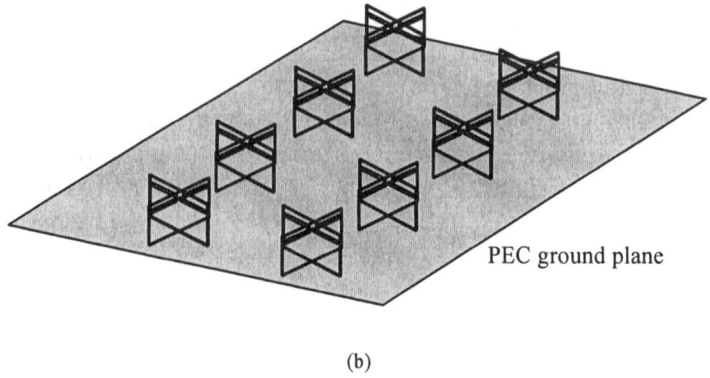

(b)

Figure 8.7 (continued.)

Following the logic used in the last subsection, we first use the parallel FDTD method to simulate a 2 × 4 antenna array comprised of eight antenna elements arranged in a rectangular lattice, which can also be simulated by the FEM and MoM whose results are used as benchmarks for comparison. The parallel FDTD and FEM codes both simulate the same structure; however, the commercial MoM code we selected cannot deal with the finite thickness of the dielectric slabs. Because of this, the result from the MoM code has a significant deviation from the parallel FDTD and FEM codes, as is evident from Figure 8.8.

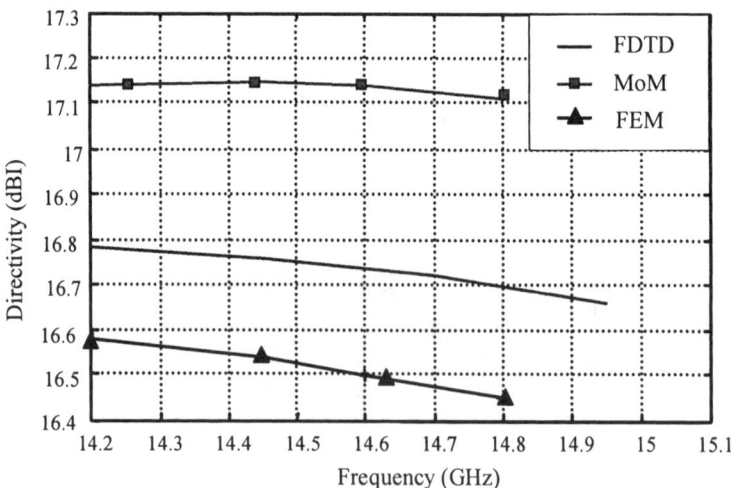

Figure 8.8 Directivity variation of the crossed dipole array with frequency.

Because a fine mesh is required to model the thin dielectric slab and the narrow width of the dipole arms, both the FEM or the MoM codes are able to handle at most 5 × 5 elements using a single processor. However, the parallel FDTD code does not suffer from such a limitation. The actual design we investigate is an array of 8 × 8 antenna elements in the rectangular lattice. The computational domain is discretized into 591 × 737 × 38 nonuniform cells. The directivity of the antenna array, computed by the parallel FDTD code, is summarized in Table 8.1.

Table 8.1
Directivity of 8 × 8 Element Antenna Array

Frequency (GHz)	Directivity (dBI)
14.0	27.15
14.3	27.25
14.7	27.36
15.0	27.47

Next, we use the crossed dipole to build up a larger antenna array that includes two units and up to 128 elements. Each unit includes 64 elements and is arranged in a rectangular lattice, as shown in Figure 8.9.

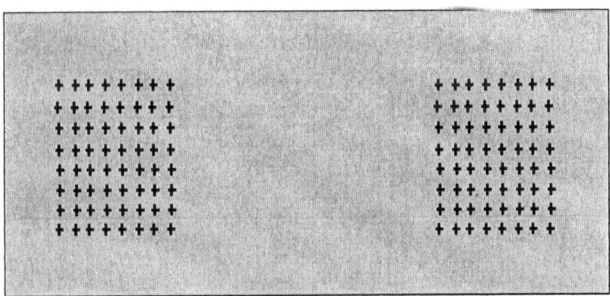

Figure 8.9 A 128-element antenna array.

For this 128-element array problem, the array size in terms of wavelength is $18.5\lambda \times 18.5\lambda \times 3.05\lambda$ and the computational domain of the FDTD simulation is discretized into 1,400 × 1,400 × 254 nonuniform cells. The thickness of the dielectric slab and the width of the dipole arms are 0.01λ and 0.1λ, respectively. We employ 60 processors in the Penn State University Lion-xm cluster to simulate this antenna array. The far-field pattern for 30° scan at 14.4 GHz is plotted in Figure 8.10.

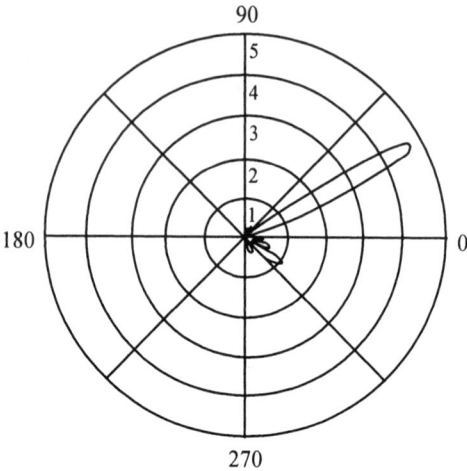

Figure 8.10 Far-field pattern of 128-element antenna array at 14.4 GHz.

The two examples described above are designed to demonstrate the capability of the parallel FDTD method to solve electrically large problems. Though not shown here, we have found that there is virtually no loss in accuracy when the code is run on a multiple processor machine versus a single one, provided that the parallelization has been carried out correctly. Although we can present more examples here, it is not really helpful for the reader's understanding of the parallel FDTD applications. Rather than presenting additional examples, we will now discuss the important steps to be followed when using the parallel FDTD code to solve practical problems, and some of the issues we need to consider before starting the simulation:

Object modeling: For objects with a canonical shape such as a cube, cylinder, and sphere, we can input the object into the computational domain by simply using their geometrical parameters. For objects with more complex shapes, we should consider how to model the objects, what software package to use for this purpose, what the output file format of the CAD modeling software should be, and how to connect it to the parallel FDTD solver.

Mesh generation: For the given problem geometry, we need to decide whether to use the uniform, the nonuniform, or the conformal meshing. The uniform mesh is quiet straightforward to use, but it is difficult to accommodate objects with arbitrary dimensions when such a mesh is used. Consequently, for most practical problems, we use a nonuniform mesh so as to be able to accurately simulate the fine features of objects, and reduce the memory requirement and enhance the simulation efficiency. To use a nonuniform mesh, we first need to find the critical points of objects. We then divide the entire computational domain into a number of subdomains such that the critical points are located at their interfaces. Finally, we design the mesh distribution inside each subdomain, using the guideline that the mesh transition from the fine to coarse cells be smooth both near the interface and in the interior region of the subdomain.

To reduce the staircasing error, it is desirable to use the conformal FDTD to simulate the problem. Although the field update equation for the conformal FDTD is relatively straightforward, the same cannot be said about the conformal mesh generation because it requires finding the intersection points between the curved surface of the object and the FDTD mesh. It is really difficult to create a robust mesh generation scheme that can cover all possible scenarios. If the conformal technique requires the knowledge of the deformed areas along the curved surface of the object, we can easily derive these areas from the intersection information. Once the information is gathered, we can use the conformal FDTD technique described in Chapter 3 to model the curved surfaces accurately.

Material distribution: The FDTD algorithm is based on the finite difference scheme, which requires a special treatment of any discontinuities of the materials. In order to calculate the electric field residing on the interface between two dissimilar materials, we need to know the parameters of an effective material that fills this area, which is calculated by using a weighted average of the adjacent materials.

Symmetric structure: It is always advisable to check if the structures being simulated as well as the excitation source are symmetric. For example, if the problem is symmetric about a line, or one, two, or three surfaces, we can use a PEC or PMC boundary condition to reduce the computational burden to be one-half, one-quarter, or one-eighth of the original problem size. If the problem is rotationally symmetric, we can reduce the problem to be a 2.5-D one.

Boundary condition: Although there exist many different types of boundary conditions, we usually need to make a choice between the PML [6] and Mur's absorbing boundary condition [7] for an open region problem. The PML provides better performance than the Mur as an ABC, but it is costlier to implement and requires more memory as well. For an engineering problem that does not require highly accurate results, Mur's absorbing boundary condition may be adequate. Neither the first-order Mur's boundary condition nor the PML requires any additional message passing in the parallel processing. For a closed region problem such as a waveguide or cavity, we usually use the PEC boundary condition to replace the PEC walls.

Excitation source: The most commonly used excitation pulse shapes are the pure or the modulated Gaussian pulse. The major difference between them is that the former has a significant DC component, and the user should make the choice for the particular problem at hand. In the FDTD simulations, a plane wave excitation can be realized either by using the scattered field formulation or the total/scattered field approach. Because of the mesh discretization approximation, the total/scattered field formulation yields a finite field distribution inside the total field region, even when there are no objects inside the computational domain. A similar phenomenon occurs in the scattered field formulation; for instance, it yields a small nonzero total field value inside PEC objects. The scattered field formulation is time consuming because it requires adding the incident field at each time step and at each point in the entire computational domain. However, they are not significantly different from the parallel processing points of view.

Output parameters: Typical output parameters are the voltages and currents in the frequency domain. We should always output the time domain signals for the derived quantities so that we can generate the result at any frequency within the band without having to rerun the FDTD code. For the far-field output, the time domain near-to-far-field transformation should be the first choice if we are only interested in the monostatic RCS; otherwise, we should consider using the frequency domain near-to-far-field transformation to reduce the memory requirement. In most applications, the far-field calculation is very time consuming; in fact, sometimes it even requires more time than the regular FDTD update. In the near-to-far-field transformation, the FFT is much faster than the DFT, but it requires a large hard disc space to store the field distribution in the time domain on the surface of Huygens' box. Therefore, we usually use the DFT to avoid the handling of a huge data file. There are two improved approaches that can be used to reduce the calculation time of the near-to-far-field transformation. The first approach is based on a down-sampling scheme in the time domain, in which we do not have to take the DFT at each FDTD time step in accordance with the sampling theorem. Since the fine features require small cell sizes, this leads to over-sampling in the time domain in the FDTD simulation. The actual sampling in the time domain is determined by the highest frequency in the spectrum of the excitation source.

The second approach is using the down-sampling in the spatial domain. The FDTD typically uses spatial over-sampling in order to describe the fine features; this over-sampling dramatically increases the burden of the near-to-far-field transformation. However, the far-field parameters are not sensitive to the current distribution, so we can use the different mesh for the far-field calculation from that used in the FDTD update. One of the practical ways is to unite some local FDTD cells on the surface of Huygens' box into one element, and then the current distribution on the united element is the weighted average of the current on the individual FDTD cells. A large number of numerical experiments have demonstrated that this simple treatment can reduce the simulation time with little compromise of the accuracy.

Parallel processing for the far-field prediction is a relatively simple task. We only need to calculate the contribution from each segment of the surface of the Huygens' box, and then obtain the final far field by using vector summation of these contributions.

Parallel design: The parallel FDTD has been discussed in previous chapters. We should consider the following factors during the parallel FDTD design to achieve a high parallel efficiency. First, we need to optimize the data exchange among the processors, in which we arrange the processor distribution according to the structure of the computational domain and the number of processors. Second, if the computational domain contains a large PEC region, we do not need to exchange any information inside the PEC region. We usually create a buffer to store the nonzero data and only exchange this information in the parallel processing. We can remove the subdomain that is embedded completely inside the PEC object. Finally, for a class of structures such as those encountered in the electronic packaging problems, the layout of the processors should be related to the number of the layers in the structures so that the data exchange between layers only is carried out in the small region around the via.

Besides the points listed above, the reader should also consider all the important factors that may affect parallel efficiency, simulation time, memory cost, and code complexity.

REFERENCES

[1] W. Yu, et al., "A Robust Parallelized Conformal Finite Difference Time Domain Field Solver Package Using the MPI Library," *IEEE Antennas and Propagation Magazine*, Vol. 47, No. 3, 2005, pp. 39-59.

[2] R. Hanington, *Field Computations by Moment Methods*, Macmillan, New York, 1968.

[3] A. Peterson, S. Ray, and R. Mittra, *Computational Methods for Electromagnetics* (IEEE Press Series on Electromagnetic Wave Theory), IEEE Press, New York, 1997.

[4] J. Jin, *The Finite Element Method in Electromagnetics*, John Wiley & Sons, New York, 2002.

[5] C. Balanis, *Antenna Theory: Analysis and Design*, 2nd ed., John Wiley & Sons, New York, 1997.

[6] J. Berenger, "A Perfectly Matched Layer Medium for the Absorption of Electromagnetic Waves," *J. Comput.*, Vol. 114, October 1994, pp. 185-200.

[7] G. Mur, "Absorbing Boundary Conditions for the Finite-Difference Approximation of the Time-Domain Electromagnetic Field Equations," *IEEE Transactions on Electromagnetic Compatibility*, Vol. 23, No. 3, 1981, pp. 377-382.

Chapter 9

FDTD Analysis of Bodies of Revolution

In this chapter we discuss the solution to problems involving objects that have a rotational symmetry, called the bodies of revolution or BORs. There are two major disadvantages to using a 3-D FDTD algorithm in cylindrical coordinates: first, there exists a singularity on the z-axis in the cylindrical FDTD formulation unless we are dealing with a shell-type of geometry; and second, for a problem geometry which has a large radius, if the cell sizes near the z-axis become very small compared to those in the region away from the axis, this can cause a high-degree numerical dispersion. However, we can take advantage of the azimuthal symmetry to reduce the original 3-D problem into a 2-D problem by factoring out the φ-variation [1–6].

In this chapter we discuss the BOR/FDTD including the basic update procedure, absorbing boundary conditions, singular boundary condition, partially symmetric problems, and near-to-far-field transformation. Also, we will wait to describe the implementation of the parallel BOR/FDTD algorithm and its engineering applications until the next chapter.

9.1 BOR/FDTD

We begin with two curl Maxwell's equations, and they can be written as:

$$\nabla \times \vec{E} = -\mu \frac{\partial \vec{H}}{\partial t} \qquad (9.1a)$$

$$\nabla \times \vec{H} = \varepsilon \frac{\partial \vec{E}}{\partial t} + \sigma \vec{E} \qquad (9.1b)$$

These equations can be expressed as the following six coupled equations in the cylindrical coordinate system:

$$\frac{\partial E_z}{\rho \partial \varphi} - \frac{\partial E_\varphi}{\partial z} = -\mu_\rho \frac{\partial H_\rho}{\partial t} \qquad (9.2\text{a})$$

$$\frac{\partial E_\rho}{\partial z} - \frac{\partial E_z}{\partial \rho} = -\mu_\varphi \frac{\partial H_\varphi}{\partial t} \qquad (9.2\text{b})$$

$$\frac{\partial (\rho E_\varphi)}{\rho \partial \rho} - \frac{\partial E_\rho}{\rho \partial \varphi} = -\mu_z \frac{\partial H_z}{\partial t} \qquad (9.2\text{c})$$

$$\frac{\partial H_z}{\rho \partial \varphi} - \frac{\partial H_\varphi}{\partial z} = \varepsilon_\rho \frac{\partial E_\rho}{\partial t} + \sigma_\rho E_\rho \qquad (9.2\text{d})$$

$$\frac{\partial H_\rho}{\partial z} - \frac{\partial H_z}{\partial \rho} = \varepsilon_\varphi \frac{\partial E_\varphi}{\partial t} + \sigma_\varphi E_\varphi \qquad (9.2\text{e})$$

$$\frac{\partial (\rho H_\varphi)}{\rho \partial \rho} - \frac{\partial H_\rho}{\rho \partial \varphi} = \varepsilon_z \frac{\partial E_z}{\partial t} + \sigma_z E_z \qquad (9.2\text{f})$$

where the field components E_ρ, E_φ, E_z, H_ρ, H_φ, and H_z are functions of the spatial variables ρ, φ, z, and the time variable t. If both the electric and magnetic fields are azimuthally symmetric, implying that they are either independent of φ or have a known φ-variation, for example, $\sin(m\varphi)$ or $\cos(m\varphi)$, then we can express these fields as though there were only functions of the spatial variables ρ, z, and the time variable t. To this end, we can rewrite (9.2a) to (9.2f) after factoring out the angular variable φ from the field. The detail of the procedure will be presented below.

9.2 UPDATE EQUATIONS FOR BOR/FDTD

The electric and magnetic fields with rotationally symmetric characteristics can be expressed as follows [1]:

$$\vec{E}(\rho,\varphi,z,t) = \sum_{m=0}^{\infty} \left[\vec{E}(\rho,z,t)_{\text{even}} \cos(m\varphi) + \vec{E}(\rho,z,t)_{\text{odd}} \sin(m\varphi) \right] \qquad (9.3)$$

$$\vec{H}(\rho,\varphi,z,t) = \sum_{m=0}^{\infty} \left[\vec{H}(\rho,z,t)_{\text{even}} \cos(m\varphi) + \vec{H}(\rho,z,t)_{\text{odd}} \sin(m\varphi) \right] \qquad (9.4)$$

where m is the mode number along the φ-direction. In writing (9.3) and (9.4), we assumed that the field equations have been expressed as:

$$E_\rho(\rho,\varphi,z,t) = \begin{pmatrix} \sin(m\varphi) \\ \cos(m\varphi) \end{pmatrix} E_\rho(\rho,z,t) \tag{9.5a}$$

$$E_\varphi(\rho,\varphi,z,t) = \begin{pmatrix} \cos(m\varphi) \\ \sin(m\varphi) \end{pmatrix} E_\varphi(\rho,z,t) \tag{9.5b}$$

$$E_z(\rho,\varphi,z,t) = \begin{pmatrix} \sin(m\varphi) \\ \cos(m\varphi) \end{pmatrix} E_z(\rho,z,t) \tag{9.5c}$$

$$H_\rho(\rho,\varphi,z,t) = \begin{pmatrix} \cos(m\varphi) \\ \sin(m\varphi) \end{pmatrix} H_\rho(\rho,z,t) \tag{9.6a}$$

$$H_\varphi(\rho,\varphi,z,t) = \begin{pmatrix} \sin(m\varphi) \\ \cos(m\varphi) \end{pmatrix} H_\varphi(\rho,z,t) \tag{9.6b}$$

$$H_z(\rho,\varphi,z,t) = \begin{pmatrix} \cos(m\varphi) \\ \sin(m\varphi) \end{pmatrix} H_z(\rho,z,t) \tag{9.6c}$$

Substituting (9.5) and (9.6) into (9.2), we obtain the following six coupled equations:

$$\pm \frac{m}{\rho} E_z - \frac{\partial E_\varphi}{\partial z} = -\mu_\rho \frac{\partial H_\rho}{\partial t} \tag{9.7a}$$

$$\frac{\partial E_\rho}{\partial z} - \frac{\partial E_z}{\partial \rho} = -\mu_\varphi \frac{\partial H_\varphi}{\partial t} \tag{9.7b}$$

$$\frac{1}{\rho}\frac{\partial}{\partial \rho}(\rho E_\varphi) \mp \frac{m}{\rho} E_\rho = -\mu_z \frac{\partial H_z}{\partial t} \tag{9.7c}$$

$$\mp \frac{m}{\rho} H_z - \frac{\partial H_\varphi}{\partial z} = \varepsilon_\rho \frac{\partial E_\rho}{\partial t} + \sigma_\rho E_\rho \tag{9.7d}$$

$$\frac{\partial H_\rho}{\partial z} - \frac{\partial H_z}{\partial \rho} = \varepsilon_\varphi \frac{\partial E_\varphi}{\partial t} + \sigma_\varphi E_\varphi \tag{9.7e}$$

$$\frac{1}{\rho}\frac{\partial(\rho H_\varphi)}{\partial \rho} \pm \frac{m}{\rho} H_\rho = \varepsilon_z \frac{\partial E_z}{\partial t} + \sigma_z E_z \tag{9.7f}$$

where each of the field components is a function of the spatial variables ρ, z, and the time variable t. The above equations are referred to as 2.5-D versions of the original equation (9.1). The locations of the electric and magnetic fields in the original 3-D and 2.5-D problems are shown in Figure 9.1.

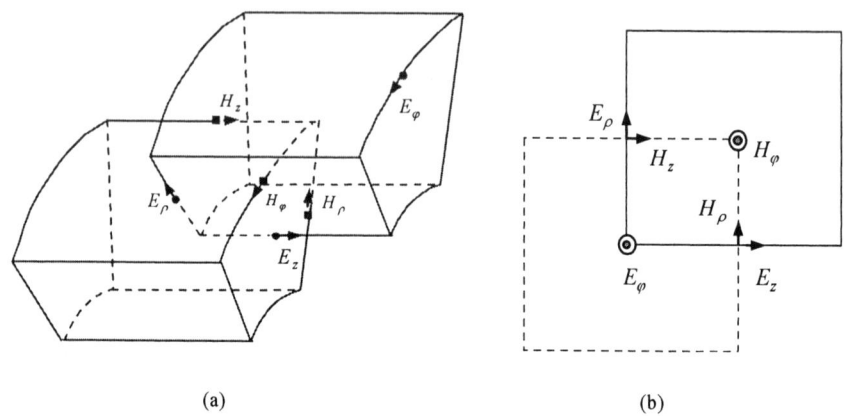

(a)   (b)

Figure 9.1 Field locations. (a) Original 3-D problem. (b) 2.5-D problem.

For the sake of simplicity, we assume that the losses in the material being simulated are nonmagnetic. Then following the same notations as those used in the 3-D FDTD, we can discretize (9.7) as:

$$H_\rho^{n+1/2}(i, j+1/2) = H_\rho^{n-1/2}(i, j+1/2) \mp \frac{m\Delta t}{\mu_\rho \rho(i)} E_z^n(i, j+1/2)$$
$$+ \frac{\Delta t}{\mu_\rho} \left[\frac{E_\varphi^n(i, j+1) - E_\varphi^n(i, j)}{\Delta z(j)} \right]$$
(9.8a)

$$H_\varphi^{n+1/2}(i+1/2, j+1/2) = H_\varphi^{n-1/2}(i+1/2, j+1/2)$$
$$- \frac{\Delta t}{\mu_\varphi \rho(i+1/2)} \left[\frac{E_\rho^n(i+1/2, j+1) - E_\rho^n(i+1/2, j)}{\Delta z(j)} \right]$$
$$+ \frac{\Delta t}{\mu_\varphi} \left[\frac{E_z^n(i+1, j+1/2) - E_z^n(i, j+1/2)}{\Delta \rho(i)} \right]$$
(9.8b)

$$H_z^{n+1/2}(i+1/2,j) = H_z^{n-1/2}(i+1/2,j)$$

$$-\frac{\Delta t}{\mu_z \rho(i+1/2)} \left[\frac{\rho(i+1)E_\varphi^n(i+1,j) - \rho(i)E_\varphi^n(i,j)}{\Delta \rho(i)} \right] \quad (9.8c)$$

$$\pm \frac{m\Delta t}{\mu_z \rho(i+1/2)} E_\rho^n(i+1/2,j)$$

$$E_\rho^{n+1}(i+1/2,j) = \frac{\varepsilon_\rho - 0.5\sigma_\rho \Delta t}{\varepsilon_\rho + 0.5\sigma_\rho \Delta t} E_\rho^n(i+1/2,j)$$

$$-\frac{\Delta t}{\varepsilon_\rho + 0.5\sigma_\rho \Delta t} \left[\frac{H_\varphi^{n+1/2}(i+1/2,j+1/2) - H_\varphi^{n+1/2}(i+1/2,j-1/2)}{\Delta z(j)} \right] \quad (9.8d)$$

$$\mp \frac{m\Delta t}{\varepsilon_\rho + 0.5\sigma_\rho \Delta t} \frac{H_z^{n+1/2}(i+1/2,j)}{\rho(i+1/2)}$$

$$E_\varphi^{n+1}(i,j) = \frac{\varepsilon_\varphi - 0.5\sigma_\varphi \Delta t}{\varepsilon_\varphi + 0.5\sigma_\varphi \Delta t} E_\varphi^n(i,j)$$

$$+\frac{\Delta t}{\varepsilon_\varphi + 0.5\sigma_\varphi \Delta t} \left[\frac{H_\rho^{n+1/2}(i+1/2,j+1/2) - H_\rho^{n+1/2}(i+1/2,j-1/2)}{\Delta z(j)} \right] \quad (9.8e)$$

$$-\frac{\Delta t}{\varepsilon_\varphi + 0.5\sigma_\varphi \Delta t} \left[\frac{H_z^{n+1/2}(i+1/2,j) - H_z^{n+1/2}(i-1/2,j)}{\Delta \rho(i)} \right]$$

$$E_z^{n+1}(i,j+1/2) = \frac{\varepsilon_z - 0.5\sigma_z \Delta t}{\varepsilon_z + 0.5\sigma_z \Delta t} E_z^n(i,j+1/2)$$

$$+\frac{\Delta t}{\varepsilon_z + 0.5\sigma_z \Delta t} \left[\frac{\rho(i+1/2) H_\varphi^{n+1/2}(i+1/2,j+1/2) - \rho(i-1/2) H_\varphi^{n+1/2}(i-1/2,j+1/2)}{\rho(i)\Delta \rho(i)} \right]$$

$$\pm \frac{m\Delta t}{\varepsilon_z + 0.5\sigma_z \Delta t} \frac{H_\rho^{n+1/2}(i,j+1/2)}{\rho(i)}$$

(9.8f)

where the plus and minus signs correspond to the choice of basis functions in (9.5) and (9.6). To handle the singularity in the above equations as $\rho \to 0$, we usually choose to align H_z with the z-axis; thus, the first cell along the ρ-direction is only a one-half cell inside the computational domain.

9.3 PML FOR BOR/FDTD

In Chapter 2 we discussed several versions of the PML, which include the original Berenger's split field PML [7], Gedney's unsplit field PML [8], and the stretched coordinate PML [9, 10]. Although these different versions of the PML are similar in principle, there are substantial differences from a programming point of view. We choose Gedney's unsplit field PML among them for implementation in the BOR/FDTD. Maxwell's equations in both the PML region and the computational domain can then be written as:

$$\nabla \times \tilde{H} = j\omega\varepsilon_0 \left(\varepsilon_r + \frac{\sigma}{j\omega\varepsilon_0}\right) \bar{\bar{\varepsilon}} \tilde{E} \tag{9.9a}$$

$$\nabla \times \tilde{E} = -j\omega\mu_0 \left(\mu_r + \frac{\sigma_M}{j\omega\mu_0}\right) \bar{\bar{\mu}} \tilde{H} \tag{9.9b}$$

where $\bar{\bar{\varepsilon}}$ and $\bar{\bar{\mu}}$ are tensors with the following expression:

$$\bar{\bar{\mu}} = \bar{\bar{\varepsilon}} = \begin{bmatrix} S_\varphi S_z S_\rho^{-1} & 0 & 0 \\ 0 & S_z S_\rho S_\varphi^{-1} & 0 \\ 0 & 0 & S_\rho S_\varphi S_z^{-1} \end{bmatrix} \tag{9.10}$$

In order to achieve a low reflection from the PML region, the parameters S_ρ, S_φ, and S_z in the above equation should be selected as follows $S_\rho = K_{\rho,PML} + \frac{\sigma_{\rho,PML}}{j\omega\varepsilon_0}$, $S_\varphi = K_{\varphi,PML} + \frac{\Sigma_{\varphi,PML}}{j\omega\varepsilon_0}$, $S_z = K_{z,PML} + \frac{\sigma_{z,PML}}{j\omega\varepsilon_0}$, where $K_{\varphi,PML}$ and $\Sigma_{\varphi,PML}$ are given by

$$K_{\varphi,PML} = \frac{\rho_0 + \int_{\rho_0}^{\rho} K_\rho(\rho')d\rho'}{\rho} \quad \text{and} \quad \Sigma_{\varphi,PML} = \frac{\int_{\rho_0}^{\rho} \sigma_\rho(\rho')d\rho'}{\rho}, \text{ where } \rho_0 \text{ is the}$$

position of the inner boundary of the PML in the ρ-direction.

Next, we will demonstrate the low reflection performance of the PML through a typical example, as shown in Figure 9.2.

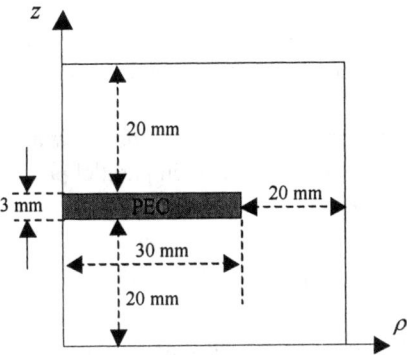

Figure 9.2 Problem configuration under test.

We will investigate the level of reflection from the PML region when the PEC disk is illuminated by a normally incident plane wave along the z-direction. The excitation source is a Gaussian pulse modulated by a sine function. Its 3-dB cutoff and modulation frequencies are 4.64 GHz and 5.2 GHz, respectively. The computational domain is discretized into 43 × 50 uniform grids and the cell size is 1 mm, which is 15 cells per wavelength at 20 GHz.

The level of reflection variation as a function of frequency from the PML region is plotted in Figure 9.3. For the sake of comparison, we also plot the reflection from the first-order Mur's boundary condition in the same figure.

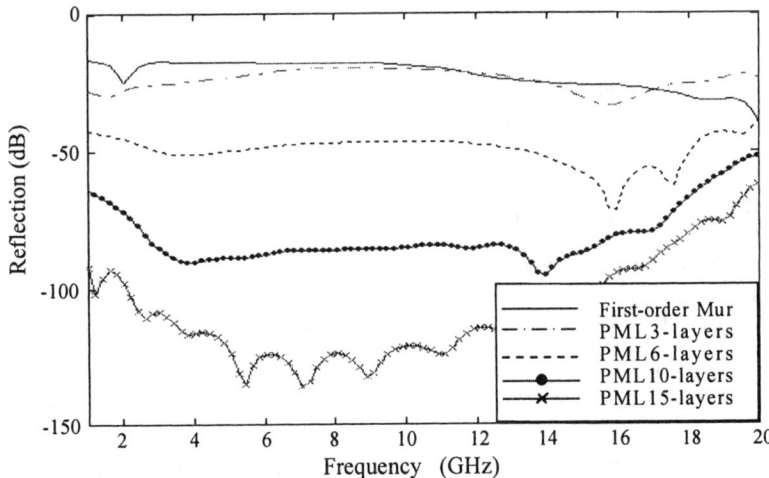

Figure 9.3 Reflection variation with frequency from the PML region.

We observe from Figure 9.3 that it is sufficient to use a six-layer PML for most engineering problems. The first-order Mur's boundary condition requires a large white space between the domain boundary and the simulated objects because its reflection level is relatively high. Although the second-order Mur's boundary condition exhibits a performance that is better (not shown here) than the first-order one, it requires extra information exchange in parallel processing, and hence, is not desirable for such processing.

9.4 NEAR-TO-FAR-FIELD TRANSFORMATION IN BOR/FDTD

Generally speaking, the near-to-far-field transformation of the BOR/FDTD is not substantially different from the 3-D FDTD Cartesian system, though the generating curve comprising Huygens' box is represented by three line segments, as shown in Figure 9.4. Consequently, we only need to take the DFT at each time step on the line segments instead of on the 3-D pillbox-type of Huygens' box, because the φ-variable can be explicitly factored out from the integral. To help the reader get a better understanding of this scheme for the far-field computation, we will elaborate on this procedure a little more below.

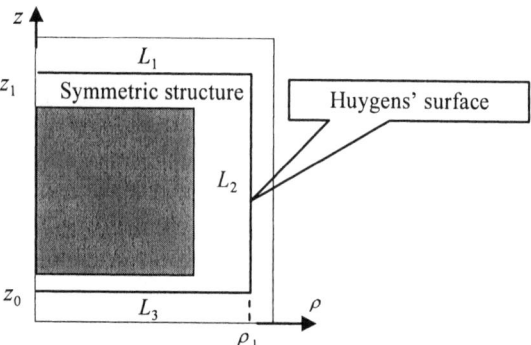

Figure 9.4 Degenerated Huygens' surface in the BOR/FDTD.

To predict the far-field contribution derived from the three line segments, L_1, L_2, and L_3, we first rotate them around the z-axis to recover the complete 3-D surface of the Huygens' box. Because we already know how the field varies with the angle φ for the different modes, it is sufficient to know the field distribution on the above three line segments to compute the far field radiated from the Huygens' box, as we detail next.

We follow the procedure used in literature [11] to calculate the far fields in the BOR/FDTD. The electric and magnetic currents in the frequency domain on the three line segments are calculated through the DFT at each time step in the BOR/FDTD simulation, and the far fields are subsequently calculated from these currents using the following expressions [12]:

$$\tilde{E}_\theta \cong -\frac{jke^{-jkr}}{4\pi r}\left(L_\varphi + \eta_0 N_\theta\right) \tag{9.11}$$

$$\tilde{E}_\varphi \cong \frac{jke^{-jkr}}{4\pi r}\left(L_\theta - \eta_0 N_\varphi\right) \tag{9.12}$$

where

$$N_\theta = \iint_S (\tilde{J}_\rho \cos\theta \cos(\varphi-\varphi') + \tilde{J}_\varphi \cos\theta \sin(\varphi-\varphi') - \tilde{J}_z \sin\theta) e^{j\beta r' \cos\psi} ds' \tag{9.13}$$

$$N_\varphi = \iint_S (-\tilde{J}_\rho \sin(\varphi-\varphi') + \tilde{J}_\varphi \cos(\varphi-\varphi')) e^{j\beta r' \cos\psi} ds' \tag{9.14}$$

$$L_\theta = \iint_S \left[\begin{array}{l}\tilde{M}_\rho \cos\theta \cos(\varphi-\varphi') \\ +\tilde{M}_\varphi \cos\theta \sin(\varphi-\varphi') - \tilde{M}_z \sin\theta\end{array}\right] e^{j\beta r' \cos\psi} ds' \tag{9.15}$$

$$L_\varphi = \iint_S (-\tilde{M}_\rho \sin(\varphi-\varphi') + \tilde{M}_\varphi \cos(\varphi-\varphi')) e^{j\beta r' \cos\psi} ds' \tag{9.16}$$

$$r'\cos\psi = \rho'\sin\theta\cos(\varphi-\varphi') + z'\cos\theta \tag{9.17}$$

where the range of integration S is a virtual, closed pillbox surface, and ($\tilde{J}_\rho, \tilde{J}_\varphi, \tilde{J}_z$) and ($\tilde{M}_\rho, \tilde{M}_\varphi, \tilde{M}_z$) are the equivalent electric and magnetic current densities on the Huygens' surface, respectively. The variables with the prime represent the coordinates of the equivalent current source. The equivalent electric and magnetic current densities on L_1, L_2, and L_3 can be expressed as:

$$\tilde{\vec{M}} = \tilde{\vec{E}} \times \hat{n} = \begin{cases} \tilde{E}_\varphi \hat{\rho} - \tilde{E}_\rho \hat{\varphi} & \text{on } L_1 \\ \tilde{E}_z \hat{\varphi} - \tilde{E}_\varphi \hat{z} & \text{on } L_2 \\ -\tilde{E}_\varphi \hat{\rho} + \tilde{E}_\rho \hat{\varphi} & \text{on } L_3 \end{cases} \tag{9.18}$$

$$\tilde{\vec{J}} = \hat{n} \times \tilde{\vec{H}} = \begin{cases} -\tilde{H}_\varphi \hat{\rho} + \tilde{H}_\rho \hat{\varphi} & \text{on } L_1 \\ -\tilde{H}_z \hat{\varphi} + \tilde{H}_\varphi \hat{z} & \text{on } L_2 \\ \tilde{H}_\varphi \hat{\rho} - \tilde{H}_\rho \hat{\varphi} & \text{on } L_3 \end{cases} \tag{9.19}$$

where $\hat{n}$ is the unit outward normal vector of the Huygens' surface. We first calculate the far field radiated by the equivalent currents on the Huygens' surface generated by the line segment L_1 rotated around the z-axis. On the line segment L_1, $z' = z_1$, $\hat{n} = \hat{z}$, $ds' = d\varphi' \rho' d\rho'$, we know from (9.18) and (9.19) that the equivalent electric and magnetic current densities are: $\tilde{J}_\rho = -\tilde{H}_\varphi$, $\tilde{J}_\varphi = \tilde{H}_\rho$, $\tilde{M}_\rho = \tilde{E}_\varphi$, and $\tilde{M}_\varphi = -\tilde{E}_\rho$. Substituting them into (9.13) to (9.16), we get:

$$N_{\theta 1} = \cos\theta e^{j\beta z_1 \cos\theta} \int_{\rho'=0}^{\rho_1} \int_{\varphi'=0}^{2\pi} \left[\begin{array}{l} -\tilde{H}_\varphi(\rho', z_1) \begin{pmatrix} \sin m\varphi' \\ \cos m\varphi' \end{pmatrix} \cos(\varphi - \varphi') \\ +\tilde{H}_\rho(\rho', z_1) \begin{pmatrix} \cos m\varphi' \\ \sin m\varphi' \end{pmatrix} \sin(\varphi - \varphi') \end{array} \right] e^{jx\cos(\varphi-\varphi')} d\varphi' \rho' d\rho' \quad (9.20)$$

$$N_{\varphi 1} = e^{j\beta z_1 \cos\theta} \int_{\rho'=0}^{\rho_1} \int_{\varphi'=0}^{2\pi} \left[\begin{array}{l} \tilde{H}_\varphi(\rho', z_1) \begin{pmatrix} \sin m\varphi' \\ \cos m\varphi' \end{pmatrix} \sin(\varphi - \varphi') \\ +\tilde{H}_\rho(\rho', z_1) \begin{pmatrix} \cos m\varphi' \\ \sin m\varphi' \end{pmatrix} \cos(\varphi - \varphi') \end{array} \right] e^{jx\cos(\varphi-\varphi')} d\varphi' \rho' d\rho' \quad (9.21)$$

$$L_{\theta 1} = \cos\theta e^{j\beta z_1 \cos\theta} \int_{\rho'=0}^{\rho_1} \int_{\varphi'=0}^{2\pi} \left[\begin{array}{l} \tilde{E}_\varphi(\rho', z_1) \begin{pmatrix} \cos m\varphi' \\ \sin m\varphi' \end{pmatrix} \cos(\varphi - \varphi') \\ -\tilde{E}_\rho(\rho', z_1) \begin{pmatrix} \sin m\varphi' \\ \cos m\varphi' \end{pmatrix} \sin(\varphi - \varphi') \end{array} \right] e^{jx\cos(\varphi-\varphi')} d\varphi' \rho' d\rho' \quad (9.22)$$

$$L_{\varphi 1} = e^{j\beta z_1 \cos\theta} \int_{\rho'=0}^{\rho_1} \int_{\varphi'=0}^{2\pi} \left[\begin{array}{l} -\tilde{E}_\varphi(\rho', z_1) \begin{pmatrix} \cos m\varphi' \\ \sin m\varphi' \end{pmatrix} \sin(\varphi - \varphi') \\ -\tilde{E}_\rho(\rho', z_1) \begin{pmatrix} \sin m\varphi' \\ \cos m\varphi' \end{pmatrix} \cos(\varphi - \varphi') \end{array} \right] e^{jx\cos(\varphi-\varphi')} d\varphi' \rho' d\rho' \quad (9.23)$$

The integration on the variable φ' in (9.20)–(9.23) can be explicitly computed without numerical calculations. Using the integral representation for the Bessel function J_m:

$$\int_0^{2\pi} e^{jm\varphi'} e^{jX\cos(\varphi-\varphi')} d\varphi' = 2\pi j^m e^{jm\varphi} J_m(X) \quad (9.24)$$

We can express the integration with respect to φ' in (9.20)–(9.23) as:

$$\int_0^{2\pi} \begin{Bmatrix} \cos(m\varphi')\cos(\varphi-\varphi') \\ \cos(m\varphi')\sin(\varphi-\varphi') \\ \sin(m\varphi')\cos(\varphi-\varphi') \\ \sin(m\varphi')\sin(\varphi-\varphi') \end{Bmatrix} e^{jX\cos(\varphi-\varphi')} d\varphi' = \begin{Bmatrix} \pi j^{m+1}\cos(m\varphi)(J_{m+1}(X)-(J_{m-1}(X)) \\ -\pi j^{m+1}\sin(m\varphi)(J_{m+1}(X)+(J_{m-1}(X)) \\ \pi j^{m+1}\sin(m\varphi)(J_{m+1}(X)-(J_{m-1}(X)) \\ \pi j^{m+1}\cos(m\varphi)(J_{m+1}(X)+(J_{m-1}(X)) \end{Bmatrix}$$

(9.25)

Inserting (9.25) into (9.20)–(9.23), we can simplify them as [11]:

$$N_{\theta 1} = \pi j^{m+1} \cos\theta e^{(j\beta z_1 \cos\theta)} \int_{\rho'=0}^{\rho_1} \left[\begin{array}{l} -\tilde{H}_\varphi \begin{Bmatrix} \sin(m\varphi) \\ \cos(m\varphi) \end{Bmatrix} (J_{m+1}(x)-(J_{m-1}(x))+ \\ \tilde{H}_\rho \begin{Bmatrix} -\sin(m\varphi) \\ \cos(m\varphi) \end{Bmatrix} (J_{m+1}(x)+(J_{m-1}(x)) \end{array} \right] \rho' d\rho' \quad (9.26)$$

$$N_{\varphi 1} = \pi j^{m+1} e^{(j\beta z_1 \cos\theta)} \int_{\rho'=0}^{\rho_1} \left[\begin{array}{l} \tilde{H}_\varphi \begin{Bmatrix} \cos(m\varphi) \\ -\sin(m\varphi) \end{Bmatrix} (J_{m+1}(x)+(J_{m-1}(x))+ \\ \tilde{H}_\rho \begin{Bmatrix} \cos(m\varphi) \\ \sin(m\varphi) \end{Bmatrix} (J_{m+1}(x)-(J_{m-1}(x)) \end{array} \right] \rho' d\rho' \quad (9.27)$$

$$L_{\theta 1} = \pi j^{m+1} \cos\theta e^{(j\beta z_1 \cos\theta)} \int_{\rho'=0}^{\rho_1} \left[\begin{array}{l} \tilde{E}_\varphi \begin{Bmatrix} \cos(m\varphi) \\ \sin(m\varphi) \end{Bmatrix} (J_{m+1}(x)-(J_{m-1}(x))- \\ \tilde{E}_\rho \begin{Bmatrix} \cos(m\varphi) \\ -\sin(m\varphi) \end{Bmatrix} (J_{m+1}(x)+(J_{m-1}(x)) \end{array} \right] \rho' d\rho' \quad (9.28)$$

$$L_{\varphi 1} = \pi j^{m+1} e^{(j\beta z_1 \cos\theta)} \int_{\rho'=0}^{\rho_1} \left[\begin{array}{l} -\tilde{E}_\varphi \begin{Bmatrix} -\sin(m\varphi) \\ \cos(m\varphi) \end{Bmatrix} (J_{m+1}(x)+(J_{m-1}(x))- \\ \tilde{E}_\rho \begin{Bmatrix} \sin(m\varphi) \\ \cos(m\varphi) \end{Bmatrix} (J_{m+1}(x)-(J_{m-1}(x)) \end{array} \right] \rho' d\rho' \quad (9.29)$$

where $x = \beta\rho'\sin\theta$. Substituting (9.26)–(9.29) into (9.11)–(9.12), we get:

$$\tilde{E}_{\theta 1} = \left(-\frac{j\beta e^{-j\beta r}}{4\pi r}\right) \pi j^{m+1} e^{(j\beta z_1 \cos\theta)} \int_{\rho'=0}^{\rho_1} \left[\begin{array}{l} (-\tilde{E}_\varphi + \eta\cos\theta\ \tilde{H}_\rho)\frac{\mp 2m}{x} J_m(x) + \\ (-\tilde{E}_\rho - \eta\cos\theta\ \tilde{H}_\varphi)(-2)J'_m(x) \end{array} \right] \rho' d\rho' \quad (9.30)$$

$$\tilde{E}_{\varphi 1} = \left(\frac{j\beta e^{-j\beta r}}{4\pi r}\right) \pi j^{m+1} e^{(j\beta z_1 \cos\theta)} \int_{\rho'=0}^{\rho_1} \left[\begin{array}{l} (\cos\theta\tilde{E}_\varphi - \eta\tilde{H}_\rho)(-2)J'_m(x) + \\ (-\cos\theta\tilde{E}_\rho - \eta\tilde{H}_\varphi)\frac{\pm 2m}{x} J_m(x) \end{array} \right] \rho' d\rho' \quad (9.31)$$

$$\tilde{E}_{\theta 1}(\rho,\varphi,z) = \begin{Bmatrix} \sin(m\varphi) \\ \cos(m\varphi) \end{Bmatrix} \tilde{E}_{\theta 1}(\rho,z) \tag{9.32}$$

$$\tilde{E}_{\varphi 1}(\rho,\varphi,z) = \begin{Bmatrix} \cos(m\varphi) \\ \sin(m\varphi) \end{Bmatrix} \tilde{E}_{\varphi 1}(\rho,z) \tag{9.33}$$

The plus and minus signs in (9.30) and (9.31) correspond to the different choice of the basis functions in (9.32) and (9.33).

Using a similar procedure, the far fields contributed from the currents on the line segment L_3 ($\hat{n} = -\hat{z}$, $z' = z_0$) can be written as:

$$\tilde{E}_{\theta 3} = -\frac{j\beta e^{-j\beta r}}{4\pi r} \pi j^{m+1} e^{j\beta z_0 \cos\theta} \int_{\rho'=0}^{\rho_1} \begin{bmatrix} (\tilde{E}_\varphi - \eta\cos\theta \tilde{H}_\rho)\frac{\mp 2m}{x} J_m(x) + \\ (\tilde{E}_\rho + \eta\cos\theta \tilde{H}_\varphi)(-2)J'_m(x) \end{bmatrix} \rho'd\rho' \tag{9.34}$$

$$\tilde{E}_{\varphi 3} = \frac{j\beta e^{-j\beta r}}{4\pi r} \pi j^{m+1} e^{j\beta z_0 \cos\theta} \times \int_{\rho'=0}^{\rho_1} \begin{bmatrix} (-\cos\theta \tilde{E}_\varphi + \eta\tilde{H}_\rho)(-2)J'_m(x) \\ +(\cos\theta \tilde{E}_\rho + \eta\tilde{H}_\varphi)\frac{\pm 2m}{x} J_m(x) \end{bmatrix} \rho'd\rho' \tag{9.35}$$

The calculating procedure for the far fields contributed from the line segment L_2 ($\rho' = \rho_1$, $\hat{n} = \hat{\rho}$ and $ds' = d\varphi'\rho'dz'$) is slightly different from those on L_1 and L_3. From (9.18) and (9.19), we have $\tilde{J}_\rho = \tilde{H}_\varphi$, $\tilde{J}_\varphi = -\tilde{H}_z$, $\tilde{M}_z = -\tilde{E}_\varphi$, and $\tilde{M}_\varphi = \tilde{E}_z$. Substituting these electric and magnetic current densities into (9.20)–(9.23), we get:

$$N_{\theta 2} = \int_{z'=z_0}^{z_1} \begin{bmatrix} -\tilde{H}_z \begin{Bmatrix} -\pi j^{m+1}\sin(m\varphi)(J_{m+1}(x)+(J_{m-1}(x)) \\ \pi j^{m+1}\cos(m\varphi)(J_{m+1}(x)+(J_{m-1}(x)) \end{Bmatrix} \cos\theta \\ -\tilde{H}_\varphi \begin{Bmatrix} 2\pi j^m J_m(x)\sin(m\varphi) \\ 2\pi j^m J_m(x)\cos(m\varphi) \end{Bmatrix} \sin\theta \end{bmatrix} e^{(j\beta z'\cos\theta)} \rho_1 dz' \tag{9.36}$$

$$N_{\varphi 2} = \int_{z'=z_0}^{z_1} -\tilde{H}_z \begin{Bmatrix} \pi j^{m+1}\cos(m\varphi)(J_{m+1}(x)-(J_{m-1}(x)) \\ \pi j^{m+1}\sin(m\varphi)(J_{m+1}(x)-(J_{m-1}(x)) \end{Bmatrix} e^{j\beta z'\cos\theta} d\varphi'\rho_1 dz' \tag{9.37}$$

$$L_{\theta 2} = \int_{z'=z_0}^{z_1} \begin{bmatrix} \tilde{E}_z \begin{Bmatrix} \pi j^{m+1}\cos(m\varphi)(J_{m+1}(x)+(J_{m-1}(x)) \\ -\pi j^{m+1}\sin(m\varphi)(J_{m+1}(x)+(J_{m-1}(x)) \end{Bmatrix} \cos\theta \\ +\tilde{E}_\varphi \begin{Bmatrix} 2\pi j^m J_m(x)\cos(m\varphi) \\ 2\pi j^m J_m(x)\sin(m\varphi) \end{Bmatrix} \sin\theta \end{bmatrix} e^{j\beta z'\cos\theta} \rho_1 dz' \tag{9.38}$$

$$L_{\varphi 2} = \int_{z'=z_0}^{z_1} \tilde{E}_z \begin{Bmatrix} \pi j^{m+1} \sin(m\varphi)(J_{m+1}(x) - J_{m-1}(x)) \\ \pi j^{m+1} \cos(m\varphi)(J_{m+1}(x) - J_{m-1}(x)) \end{Bmatrix} e^{j\beta z'\cos\theta} \rho_1 dz' \quad (9.39)$$

where $x = \beta\rho_1 \sin\theta$. Substituting (9.36)–(9.39) into (9.11)–(9.12), we get:

$$\tilde{E}_{\theta 2} = \left(-\frac{j\beta \exp(-j\beta r)}{4\pi r}\right) 2\pi j^m \begin{bmatrix} J_m'(x) \int_{z'=z_0}^{z_1} \rho_1 j\tilde{E}_z e^{(j\beta z'\cos\theta)} dz' + \\ \eta J_m(x) \int_{z'=z_0}^{z_1} (\tilde{H}_z j \frac{\mp m}{\beta_m} - \tilde{H}_\varphi \rho_1 \sin\theta) e^{(j\beta_m z'\cos\theta)} dz' \end{bmatrix} \quad (9.40)$$

$$\tilde{E}_{\varphi 2} = \frac{j\beta \exp(-j\beta r)}{4\pi r} 2\pi j^m \begin{bmatrix} J_m(x) \int_{z'=z_0}^{z_1} (\tilde{E}_z j \frac{\pm m}{\beta_m} + \tilde{E}_\varphi \rho_1 \sin\theta) e^{j\beta z'\cos\theta} dz' \\ -\eta J_m'(x) \int_{z'=z_0}^{z_1} \rho_1 j\tilde{H}_z e^{(j\beta z'\cos\theta)} dz' \end{bmatrix} \quad (9.41)$$

Finally, the total far fields contributed by the equivalent electric and magnetic currents on the line segments L_1, L_2, and L_3 are computed from:

$$\tilde{E}_\theta = \tilde{E}_{\theta 1} + \tilde{E}_{\theta 2} + \tilde{E}_{\theta 3} \quad (9.42)$$

$$\tilde{E}_\varphi = \tilde{E}_{\varphi 1} + \tilde{E}_{\varphi 2} + \tilde{E}_{\varphi 3} \quad (9.43)$$

Next, to illustrate the procedure outlined above we apply it to calculate the far fields of feed systems of a parabolic reflector antenna [11]. The configuration of the system is shown in Figure 9.5.

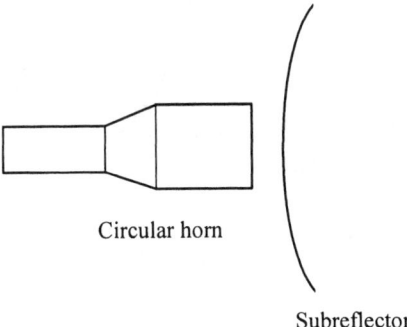

Figure 9.5 Feed system of a parabolic reflector antenna.

The far-field pattern of the parabolic reflector antenna is plotted in Figure 9.6. For the sake of comparison, we also plot the result calculated using the physical optics (PO) in the same figure. Good agreement between the BOR/FDTD and the PO method is evident from the plots.

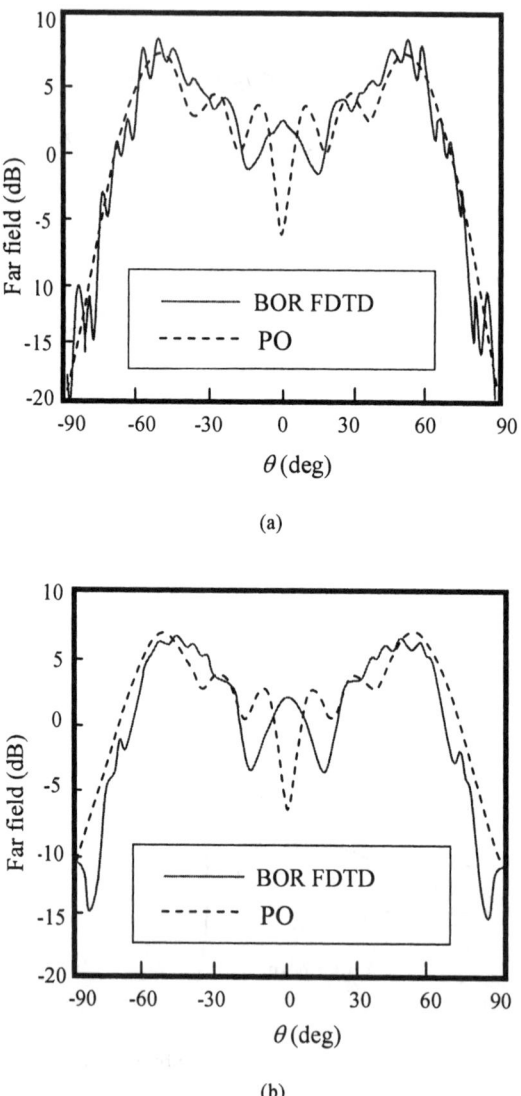

Figure 9.6 Far-field pattern of a feed system. (a) Far field in E-plane. (b) Far field in H-plane.

9.5 SINGULAR BOUNDARY IN THE BOR/FDTD

Since only one-half of the first cell $\Delta\rho(0)$ in the radial direction is located in the computational domain, only the magnetic fields H_z and H_φ are on the z-axis. We will now discuss the treatment of the singular property of these fields as $\rho \to 0$. The singular characteristics of the field components preclude the use of the update equations (9.8a)–(9.8f) on the z-axis. However, the magnetic fields are associated with the electric fields around the z-axis via Faraday's law [2], and using them we can calculate these magnetic fields on the z-axis in (9.8c). We need to know $H_z^{n+1/2}(-1/2, j)\big|_{\rho=0}$ on the z-axis in order to calculate $E_\varphi^{n+1}(0, j)\big|_{\rho=0.5\Delta\rho(0)}$. The integral formulation of Faraday's law has the form:

$$\oint_{\Delta c} \vec{E} \cdot d\vec{l} = -\iint_{\Delta s} \mu \frac{\partial \vec{H}}{\partial t} \cdot d\vec{s} \qquad (9.44)$$

where Δc is loops around the z-axis and Δs is an area inside the loop, as shown in Figure 9.7(a).

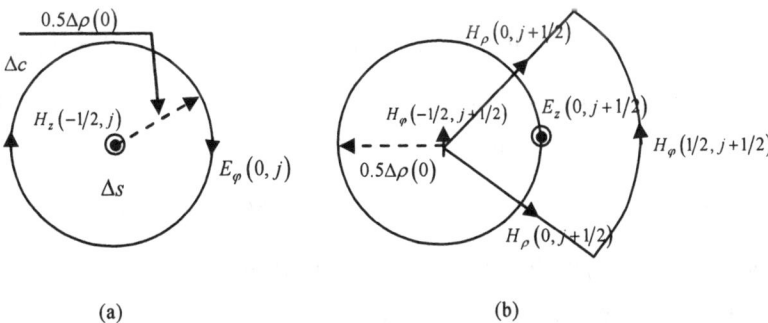

(a) (b)

Figure 9.7 Singularity treatment of the magnetic field on the z-axis. (a) Treatment of $H_z^{n+1/2}$ on the z-axis. (b) Treatment of E_z^{n+1} on the z-axis.

We can next apply (9.44) to the magnetic field H_z on the z-axis, and the electric field E_φ around the same. Since the electric field E_φ is either a $\sin(\varphi)$ or $\cos(\varphi)$ variation, for $m > 0$ we have:

$$0.5\Delta\rho(0) \int_0^{2\pi} E_\varphi \cos(m\varphi) \, d\varphi = 0 \qquad (9.45)$$

$$0.5\Delta\rho(0) \int_0^{2\pi} E_\varphi \sin(m\varphi) \, d\varphi = 0 \qquad (9.46)$$

It follows then, that $H_z^{n+1/2}(-1/2, j)\big|_{\rho=0}$ is always equal to zero for $m > 0$. If the mode number $m = 0$, the electric field $E_\varphi^n(0, j)\big|_{\rho=0.5\Delta\rho(0)}$ is a constant, so (9.44) becomes:

$$2\pi\left[0.5\Delta\rho(0)\right]E_\varphi^n = -\mu_z\pi\left[0.5\Delta\rho(0)\right]^2 \frac{H_z^{n+1/2} - H_z^{n-1/2}}{\Delta t} \qquad (9.47)$$

Simplifying (9.47) we get:

$$H_z^{n+1/2} = H_z^{n-1/2} - \frac{4\Delta t}{\mu_z \Delta\rho(0)} E_\varphi^n \qquad (9.48)$$

Once we have obtained $H_z^{n+1/2}(-1/2, j)\big|_{\rho=0}$ from (9.48), we can calculate $E_\varphi^{n+1}(0, j)\big|_{\rho=0.5\Delta\rho(0)}$ from the BOR/FDTD update equations. Examining the update equation (9.8f) we find that we need to know the magnetic field $H_\varphi^{n+1/2}(-1/2, j+1/2)\big|_{\rho=0}$ on the z-axis to calculate $E_z^{n+1}(0, j+1/2)\big|_{\rho=0.5\Delta\rho(0)}$, as shown in Figure 9.7. Because the magnetic field $H_\varphi^{n+1/2}(-1/2, j+1/2)\big|_{\rho=0}$ has a multiplicative factor $\rho(-1/2) = 0$, the magnetic field $H_\varphi^{n+1/2}(-1/2, j+1/2)$ has no contribution to the electric field calculation in (9.8f).

9.6 PARTIALLY SYMMETRIC PROBLEM

In many practical applications involving BOR geometries, the excitation source is not fully symmetric about the z-axis, for example, neither the normally nor the obliquely incident plane waves are φ-symmetric. Although we cannot directly apply the BOR/FDTD to simulate these asymmetric problems, we are still able to use it if we represent the excitation source in a slightly different way [3].

9.6.1 Normally Incident Plane Wave

In order to simulate a normally incident plane wave using the BOR/FDTD algorithm, we need to split an incident plane wave into the summation of E_ρ and E_φ in the cylindrical coordinate system. For instance, a plane wave with the polarization of E_x, shown in Figure 9.8(a), can be expressed as:

$$\hat{x} E_x = \hat{\rho} E_\rho - \hat{\varphi} E_\varphi \qquad (9.49)$$

while E_ρ and E_φ can be expressed as $E_x \cos(\varphi)$ and $-E_x \sin(\varphi)$, respectively. Likewise, a plane wave with the polarization of E_y, shown in Figure 9.8(b), can be expressed as:

$$\hat{y} E_y = \hat{\rho} E_\rho + \hat{\varphi} E_\varphi \qquad (9.50)$$

while E_ρ and E_φ can be expressed as $E_y \sin(\varphi)$ and $E_y \cos(\varphi)$, respectively.

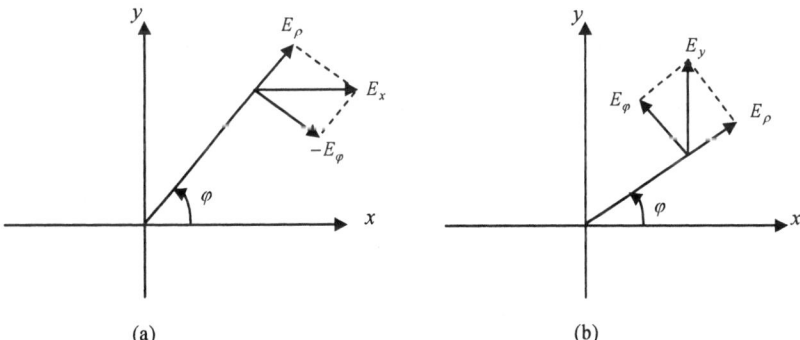

Figure 9.8 Simulation of a normally incident plane wave. (a) E_x polarization. (b) E_y polarization.

A plane wave with the polarization of E_x in the update equations can be expressed as:

$$\frac{\partial(H_z \sin(\varphi))}{\rho \partial \varphi} - \frac{\partial(H_\varphi \cos(\varphi))}{\partial z} = \varepsilon_\rho \frac{\partial(E_\rho \cos(\varphi))}{\partial t} + (\sigma_\rho E_\rho + E_x)\cos(\varphi) \quad (9.51)$$

$$\frac{\partial(H_\rho \sin(\varphi))}{\partial z} - \frac{\partial(H_z \sin(\varphi))}{\partial \rho} = \varepsilon_\phi \frac{\partial(H_\varphi \sin(\varphi))}{\partial t} + (\sigma_\varphi E_\varphi - E_x)\sin(\varphi) \quad (9.52)$$

Similarly, a plane wave with the polarization of E_y can be simulated by using two rotationally symmetric excitations E_ρ and E_φ, and the corresponding update equations can be expressed as:

$$\frac{\partial(H_z \cos(\varphi))}{\rho \partial \varphi} - \frac{\partial(H_\varphi \sin(\varphi))}{\partial z} = \varepsilon_\rho \frac{\partial(E_\rho \sin(\varphi))}{\partial t} + (\sigma_\rho E_\rho + E_y)\sin(\varphi) \quad (9.53)$$

$$\frac{\partial(H_\rho \cos(\varphi))}{\partial z} - \frac{\partial(H_z \cos(\varphi))}{\partial \rho} = \varepsilon_\varphi \frac{\partial(H_\varphi \cos(\varphi))}{\partial t} + (\sigma_\varphi E_\varphi + E_y)\cos(\varphi) \quad (9.54)$$

It is worthwhile noting that we assume the plane wave to be propagating along the z-axis.

9.6.2 Obliquely Incident Plane Wave

For an obliquely incident plane wave, we cannot simply resolve it into a summation of two rotationally symmetric excitations. We will now introduce an approach that can be applied to represent an obliquely incident plane wave as a summation of a series of rotationally symmetric excitations, as shown in Figure 9.9.

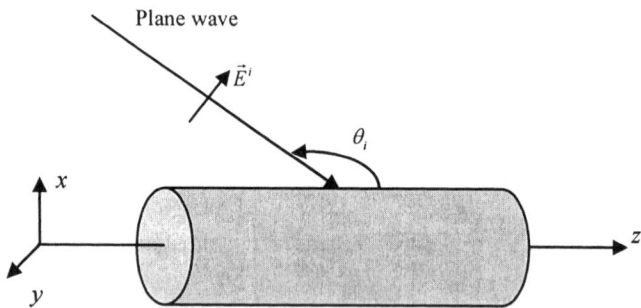

Figure 9.9 An obliquely incident plane wave.

Generally speaking, we can represent any wave in a homogenous lossless medium in terms of the summation of plane waves. For instance, for an obliquely incident plane wave given by [12]:

$$\vec{E}^i = \vec{E}_0(\hat{x}\cos\theta_i + \hat{z}\sin\theta_i)\, e^{j(-kx\sin\theta_i - kz\cos\theta_i) - j\omega t} \qquad (9.55)$$

we can express the two components of the electric field using the Bessel functions as:

$$E_x^i \cong E_0 \cos\theta_i e^{-jkz\cos\theta_i}\left(J_0(k\rho\sin\theta_i) + 2\sum_{n=1}^{N} j^{-n} J_n(k\rho\sin\theta_i)\cos(n\varphi)\right) \qquad (9.56)$$

$$E_z^i \cong E_0 \sin\theta_i e^{-jkz\cos\theta_i}\left(J_0(k\rho\sin\theta_i) + 2\sum_{n=1}^{N} j^{-n} J_n(k\rho\sin\theta_i)\cos(n\varphi)\right) \qquad (9.57)$$

where the maximum index N is determined by the incident angle θ_i, as explained below. Although the original source given in (9.55) is not rotationally symmetric about the z-axis, its two components expressed in (9.56) and (9.57) are rotationally symmetric and are explicitly expressed in terms of φ. From the asymptotic behavior of the Bessel function for a large argument, we know that they decay rapidly once the order slightly exceeds the argument. Using this property we can estimate the upper bound of N using the following formula:

$$N = k\rho_{\max} \sin\theta_i + 6 \qquad (9.58)$$

where $\rho_{\max}$ is the maximum radius of the object being simulated.

As mentioned before, the field component E_x^i can be expressed as the summation of E_ρ and $-E_\varphi$ in the cylindrical coordinate system; that is, we can write $E_\rho = E_x^i \cos(\varphi)$ and $E_\varphi = -E_x^i \sin(\varphi)$. We know from (9.54) that when $n \geq 1$, E_x^i is a function of $\cos(n\varphi)$. Therefore, the rotationally symmetric field components in the cylindrical coordinate system can be expressed as:

$$E_\rho \propto \cos(n\varphi)\cos(\varphi) \qquad (9.59a)$$

$$E_\rho \propto -\cos(n\varphi)\sin(\varphi) \qquad (9.59b)$$

It is straightforward to rewrite the products appearing in (9.59) as two-term summations of sine and cosine functions; we have:

$$\cos(n\varphi)\cos(\varphi) = \cos\big[(n+1)\varphi\big] + \cos\big[(n-1)\varphi\big] \qquad (9.60)$$

Thus, observing from (9.60), we need to simulate the BOR/FDTD code twice to calculate the fields produced by a source that varies as $\cos(n\varphi)$, and whose E_ρ consists of two components $\cos[(n+1)\varphi]$ and $\cos[(n-1)\varphi]$. The relevant equations corresponding to the above two sources can be written as [3]:

$$\frac{\partial[H_z \sin(m\varphi)]}{\rho \partial \varphi} - \frac{\partial[H_\varphi \cos(m\varphi)]}{\partial z} = \varepsilon \frac{\partial[E_\rho \cos(m\varphi)]}{\partial t} + \frac{1}{2} E_{x0}(\theta_i) \cos[(n+1)\varphi] \quad (9.61)$$

$$\frac{\partial[H_z \sin(m\varphi)]}{\rho \partial \varphi} - \frac{\partial[H_\varphi \cos(m\varphi)]}{\partial z} = \varepsilon \frac{\partial[E_\rho \cos(m\varphi)]}{\partial t} + \frac{1}{2} E_{x0}(\theta_i) \cos[(n-1)\varphi] \quad (9.62)$$

where the mode number m in (9.61) and (9.62) should be chosen to be $(n+1)$ and $(n-1)$, respectively. Finally, the solution to an obliquely incident plane wave problem is constructed by summing up the contributions of all the N modes.

Although we can apply the BOR/FDTD algorithm to simulate a rotationally symmetric structure under a plane wave incidence, we should realize that the simulation time dramatically increases with an increase in the incident angle, and can even become impractical for large oblique angles. This is because the maximum index N in (9.56) and (9.57) increases significantly with the incident angle, and each mode requires simulating the BOR/FDTD twice for mode numbers $n \geq 1$. In addition, the time step in the BOR/FDTD is a function of the mode number, which decreases rapidly with an increase of the mode number. The relationship between the time step and the mode number can be written [1]:

$$\Delta t \leq \begin{cases} \dfrac{1}{(m+1)c\sqrt{\left(\dfrac{1}{\Delta \rho}\right)^2 + \left(\dfrac{1}{\Delta z}\right)^2}}, & m > 0 \\[2ex] \dfrac{1}{\sqrt{2}c\sqrt{\left(\dfrac{1}{\Delta \rho}\right)^2 + \left(\dfrac{1}{\Delta z}\right)^2}}, & m = 0 \end{cases} \quad (9.63)$$

where c is velocity of light in free space. There is no known way to circumvent this inherent problem in the BOR formulation. All we can do is to use a parallel version of the code to reduce the run times either by using the domain decomposition approach or by running different modes on different processors. The parallelization of BOR/FDTD is the subject of discussion of the next chapter.

REFERENCES

[1] A. Taflove, *Computational Electromagnetics: The Finite-Difference Time-Domain Method*, Artech House, Norwood, MA, 2000.

[2] Y. Chen and R. Mittra, "Finite-Difference Time-Domain Algorithm for Solving Maxwell's Equations in Rotationally Symmetric Geometries," *IEEE Transactions on Microwave Theory and Techniques*, Vol. 44, No. 6, June 1996, pp. 832-839.

[3] W. Yu, D. Arakaki, and R. Mittra, "On the Solution of a Class of Large Body Problems with Full or Partial Circular Symmetry by Using the Finite Difference Time Domain (FDTD) Method," *IEEE Transactions on Antennas and Propagation*, Vol. 48, No. 12, December 2000, pp. 1810-1817.

[4] W. Yu, N. Farahat, and R. Mittra, "Far Field Pattern Calculation in BOR-FDTD Method," *Microwave and Optical Technology Letters*, Vol. 31, No. 1, October 2001, pp. 47-50.

[5] D. Arakaki, W. Yu, and R. Mittra, "On the Solution of a Class of Large Body Problems with Partial Circular Symmetry (Multiple Asymmetries) by Using a Hybrid Dimensional Finite Difference Time Domain (FDTD) Method," *IEEE Transactions on Antennas and Propagation*, Vol. 49, No. 3, March 2001, pp. 354-360.

[6] Z. Yang, et al., "Hybrid FDTD/AUTOCAD Method for the Analysis of BOR Horns and Reflectors," *Microwave and Optical Technology Letters*, Vol. 37, No. 4, May 2003, pp. 236-243.

[7] J. Berenger, "A Perfectly Matched Layer Medium for the Absorption of Electromagnetic Waves," *J. Computations*, Vol. 114, October 1994, pp. 185-200.

[8] S. Gedney, "An Anisotropic Perfectly Matched Layer-Absorbing Medium for the Truncation of FDTD Lattices," *IEEE Transactions on Antennas and Propagation*, Vol. 44, No. 12, December 1996, pp. 1630-1639.

[9] W. Chew and W. Wood, "A 3-D Perfectly Matched Medium from Modified Maxwell's Equations with Stretched Coordinates," *Microwave and Optical Technology Letters*, Vol. 7, 1994, pp. 599-604.

[10] C. Rapaport, "Perfectly Matched Absorbing Boundary Conditions Based on Anisotropic Lossy Mapping of Space," *IEEE Microwave and Guided Wave Letters*, Vol. 5, 1995, pp. 90-92.

[11] N. Farahat, W. Yu, and R. Mittra, "A Fast Near-to-Far Field Transformation Technique in BOR-FDTD Method," *IEEE Transactions on Antennas and Propagation*, Vol. 51, No. 9, September 2003, pp. 2534-2540.

[12] C. Balanis, *Advanced Engineering Electromagnetics*, John Wiley & Sons, New York, 1989, pp. 614-618.

Chapter 10

Parallel BOR/FDTD

As we know, the BOR/FDTD can be used to solve electrically large problems that are rotationally symmetric. As pointed out in the last chapter, the BOR/FDTD formulation requires simulations that are very time consuming for large bodies, especially when the angle of incidence of the impinging plane wave source is large. This prompts us to examine ways to reduce the simulation time by using parallelization, which will be discussed below.

10.1 INTRODUCTION TO PARALLEL BOR/FDTD

Parallel BOR/FDTD is designed to simulate a large task using a computer cluster in which each processor is assigned a part of the job [1–3]. Since these jobs are not independent, the processors need to share necessary information with their neighbors to update the fields at the subdomain interfaces. The exchanged fields adjacent to the subdomain interface have roles similar to the regular boundary condition. The treatment of the excitation source and the result collection follow the same procedure as described in the 3-D parallel FDTD method. However, the actual code structure is much simpler than that in the 3-D case. Although the presence of the singularity near the axis is a unique feature of the BOR/FDTD method that requires special attention, this is not really a cause for concern insofar as the parallelization is concerned because the interface between subdomains is never placed on the z-axis, and, hence, the singular boundary condition truncated by the subdomain interface does not require transferring any information in the parallel processing.

In common with the 3-D case, we still utilize an overlapping region to implement the magnetic field exchange between the adjacent subdomains. Below, we use a 2-D parallel approach to explaining the implementation of the parallel BOR/FDTD algorithm, as depicted in Figure 10.1.

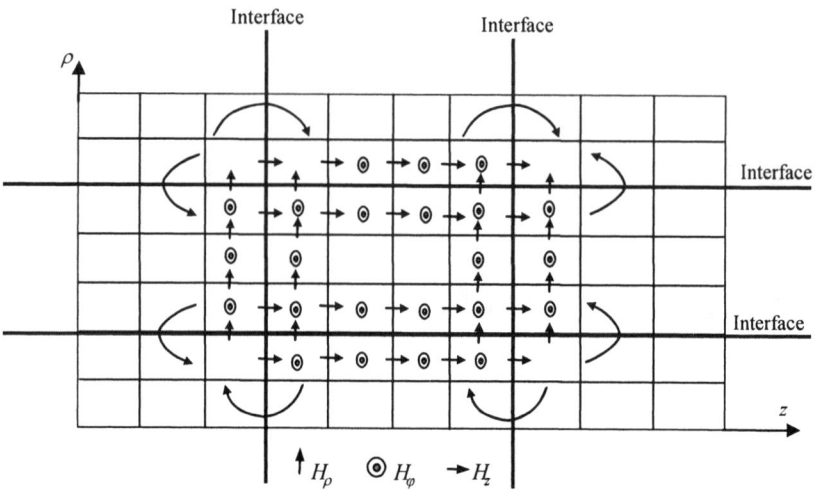

Figure 10.1 Exchanging procedure of magnetic fields in 2-D parallel BOR/FDTD.

We observe from Figure 10.1 that one of the subdomains that does not touch the global domain boundary will require the exchange of magnetic fields with the four neighboring subdomains. However, each subdomain will not exchange information with the subdomains attached at the corner. The subdomains that include the global domain boundary do not require any special treatment regardless of whether the PEC, PMC, the first-order Mur, or the PML boundary is used.

In common with the 3-D parallel case, if an excitation source or result output region is completely located within a single subdomain, the procedure is the same as that in the serial code. However, if an excitation resides on an interface between two subdomains, we need to include the excitation on the interface of both sides. In contrast to the excitation source, the output path of the result on the interface should only be taken into account at one side. Before closing this section, we would like to mention that each processor runs the same code regardless of the distribution of the processors, implying that the same variables are defined in each processor, regardless of whether it includes an excitation or output.

10.2 IMPLEMENTATION OF PARALLEL BOR/FDTD

In this section, we will present a demo of 2-D parallel BOR/FDTD code. Once again, the codes in this book are used for the demonstration purpose only, and they are not optimized. The MPI functions used in this code segment have been described in the 3-D parallel FDTD code in Chapter 7, and we will not interpret them again in this section. The reader may check Appendix A for more MPI functions. The 2-D parallel BOR/FDTD demo code in the C programming language corresponding to Figure 10.1 is listed here:

Code segment 10.1
```
VOID EXCHANGE_H_FIELD( ) {
int tag, count, i;
MPI_STATUS  status[16];
MPI_REQUEST  request[16];
for(i=0; i < exchInfo.Neighbor_Num; i++ ) {
if(exchInfo.Neighbor[i].NPos == rouLeft)         {
count = g_num_max_z;
tag=0;
MPI_ISEND(&g_hz[1][0] , count, MPI_DOUBLE, exchInfo.Neighbor[i].index,
             tag,MPI_COMM_WORLD,&request[4]);
tag=1;
MPI_IRECV(&g_hz[0][0], count, MPI_DOUBLE, exchInfo.Neighbor[i].index, tag,
             MPI_COMM_WORLD, &request[5]);
tag=2;
MPI_ISEND(&g_hphi[1][0], count, MPI_DOUBLE,
             exchInfo.Neighbor[i].index,tag,MPI_COMM_WORLD,&request[6]);
tag=3;
MPI_IRECV(&g_hphi[0][0], count, MPI_DOUBLE, exchInfo.Neighbor[i].index,
             tag,MPI_COMM_WORLD, &request[7]);
MPI_WAIT (&request[4], &status[4] ) ,
MPI_WAIT (&request[5], &status[5] ) ;
MPI_WAIT (&request[6], &status[6] ) ;
MPI_WAIT (&request[7], &status[7] ) ;        }
if(exchInfo.Neighbor[i].NPos == rouRight )  {
count = g_num_max_z;
tag=0;
MPI_IRECV(&g_hz[g_num_max_r-1][0], count, MPI_DOUBLE,
          exchInfo.Neighbor[i].index, tag, MPI_COMM_WORLD, &request[0]);
tag=1;
MPI_ISEND(&g_hz[g_num_max_r-2][0] , count, MPI_DOUBLE,
          exchInfo.Neighbor[i].index, tag, MPI_COMM_WORLD, &request[1]);
tag=2;
MPI_IRECV(&g_hphi[g_num_max_r-1][0], count, MPI_DOUBLE,
          exchInfo.Neighbor[i].index, tag, MPI_COMM_WORLD, &request[2]);
tag=3;
MPI_ISEND(&g_hphi[g_num_max_r-2][0], count, MPI_DOUBLE,
          exchInfo.Neighbor[i].index, tag, MPI_COMM_WORLD, &request[3]);
MPI_WAIT (&request[0], &status[0]) ;
```

```
MPI_WAIT (&request[1], &status[1]) ;
MPI_WAIT (&request[2], &status[2]) ;
MPI_WAIT (&request[3], &status[3]) ;   }
if(exchInfo.Neighbor[i].NPos == zUp ) {
count=1;
tag=0;
MPI_IRECV(&g_hrho[0][g_num_max_z-1], count, NewType,
          exchInfo.Neighbor[i].index, tag, MPI_COMM_WORLD, &request[12]);
tag=1;
MPI_ISEND(&g_hrho[0][g_num_max_z-2], count, NewType,
          exchInfo.Neighbor[i].index, tag, MPI_COMM_WORLD,&request[13]);
tag=2;
MPI_IRECV(&g_hphi[0][g_num_max_z-1], count, NewType,
          exchInfo.Neighbor[i].index, tag, MPI_COMM_WORLD, &request[14]);
tag=3;
MPI_ISEND(&g_hphi[0][g_num_max_z-2], count, NewType,
          exchInfo.Neighbor[i].index, tag, MPI_COMM_WORLD, &request[15]);
MPI_WAIT (&request[12], &status[12] );
MPI_WAIT (&request[13], &status[13] );
MPI_WAIT (&request[14], &status[14] );
MPI_WAIT (&request[15], &status[15] );  }
if( exchInfo.Neighbor[i].NPos == zDown ) {
count=1;
tag=0;
MPI_ISEND(&g_hrho[0][1], count, NewType, exchInfo.Neighbor[i].index, tag,
          MPI_COMM_WORLD, &request[8]);
tag=1;
MPI_IRECV(&g_hrho[0][0], count, NewType, exchInfo.Neighbor[i].index, tag,
          MPI_COMM_WORLD, &request[9]);
tag=2;
MPI_ISEND(&g_hphi[0][1], count, NewType, exchInfo.Neighbor[i].index, tag,
          MPI_COMM_WORLD, &request[10]);
tag=3;
MPI_IRECV(&g_hphi[0][0], count, NewType, exchInfo.Neighbor[i].index, tag,
          MPI_COMM_WORLD, &request[11]);
MPI_WAIT (&request[8], &status[8]);
MPI_WAIT (&request[9], &status[9]);
MPI_WAIT (&request[10], &status[10]);
MPI_WAIT (&request[11], &status[11]);  }    }
MPI_BARRIER (MPI_COMM_WORLD);  }
```

The code segment above has been successfully compiled and run on multiple platforms such as Windows, Linux, Unix, and SGI. Here, it is worthwhile mentioning that a 2-D array structure in the Fortran programming language is shown in Figure 10.2.

3	6	9	12	15	18	21	24	27
2	5	8	11	14	17	20	23	26
1	4	7	10	13	16	19	22	25

Figure 10.2 A 2-D array structure in the Fortran programming language.

We can see from Figure 10.2 that the elements of a 2-D array in the Fortran programming language are continuous in the ρ-direction. The element distribution of a 2-D array in the C programming language is shown in Figure 10.3. We have observed, however, that the elements of a 2-D array in the C programming language are continuous in the z-direction. In order to reach the best code performance, the second variable should be placed in the outer loop in the Fortran programming language, while the first variable should be placed in the outer loop in the C programming language. The variable order in the loop statement may result in a 20% difference of the code performance.

19	20	21	22	23	24	25	26	27
10	11	12	13	14	15	16	17	18
1	2	3	4	5	6	7	8	9

Figure 10.3 A 2-D array structure in the C programming language.

If we simply allocate the memory for a 2-D array in the C programming language in the following format:

ez = new double *[nrho + 1];
for (i = 0; i <= nrho; i++)   ez = new double ez[i][nz + 1];

We cannot calculate the distance from one dimension to another, even when each dimension of a 2-D array is continuous in the physical memory. In order to ensure that the physical memory location of a 2-D array is continuous, we usually use the following approach to allocating a 2-D array in the C programming language:

```
ez = new ( nothrow ) double *[nrho + 1] ;
ez[0] = new ( nothrow ) double[(nrho + 1) * (nz + 1)] ;
for ( i = 1 ; i <= nrho ; i++ )  ez[i] = ez[0] + i * (nz + 1) ;
```

10.3 EFFICIENCY ANALYSIS OF PARALLEL BOR/FDTD

In this section, we analyze the efficiency of parallel BOR/FDTD code through a simple example. A rationally symmetric reflector antenna whose radius is 160 wavelengths is located in free space and the computational domain is discretized into 3,408 × 1,632 uniform cells (20 cells per wavelength). We first test the efficiency of parallel BOR/FDTD code on an SGI workstation (shared memory system). The efficiency of parallel BOR/FDTD code is plotted in Figure 10.4. A 1-D parallel configuration in Figure 10.4 has 1, 2, 4, 6, 8, and 12 processors along the ρ- and z-directions, respectively, and a 2-D parallel configuration has processor distributions 1×1, 2×2, 3×2, 4×2, and 4×3, respectively. For a shared memory system, since the data exchange is carried out interior of the memory, its efficiency is rather high compared to a distributed memory system. It has been thematically proven that if the processors are distributed along the z-direction, the efficiency is higher than the case in which the processors are distributed along the ρ-direction. However, we have not observed much difference between the processors along the ρ- and z-directions because this system has a very high efficiency for both cases.

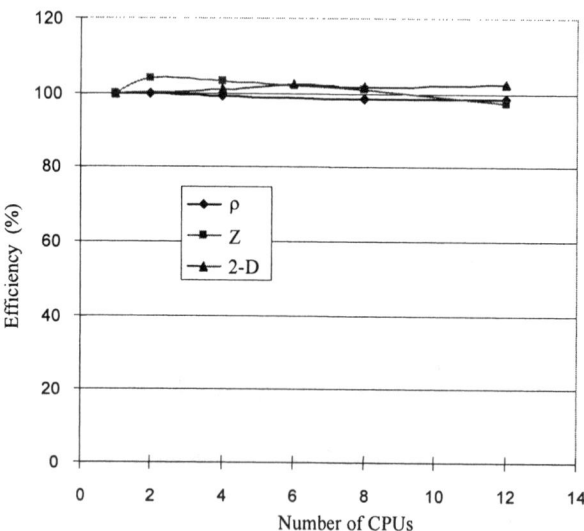

Figure 10.4 Efficiency of the parallel BOR/FDTD on a shared memory system.

Penn State Lion-xm cluster is a distributed memory system, and the efficiency of parallel BOR/FDTD code on this cluster is shown in Figure 10.5. In a 1-D parallel test, the processors are distributed as 1, 4, 8, 12, 16, 24, and 32 along the ρ- and z-directions, respectively, and 1×1, 2×2, 4×2, 4×3, 4×4, 6×4, and 8×4 in the 2-D parallel test. Compared to the shared memory system, a distributed memory system has a much lower efficiency. The unexpected variation in Figure 10.5 may be caused by instability of the cluster. The Penn State Lion-xm system has 180 nodes (360 processors), and each node has 4 GB of memory. In order to avoid that the parallel BOR/FDTD code share a node with other application codes, we always assign the even number of processors to the BOR/FDTD code.

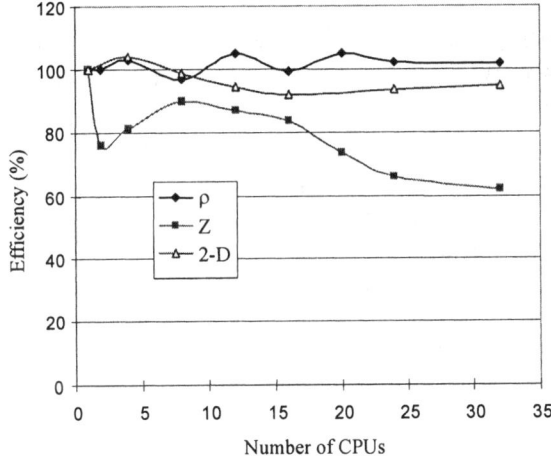

Figure 10.5 Parallel efficiency of the BOR/FDTD code on a distributed memory system.

In the rest of this chapter, we discuss the application of the parallel BOR/FDTD method to simulation of a reflector antenna system with a partially symmetric feed system. Provided that we follow the procedure described in this book, we only need to input the problem geometry and specify the excitation source, output options, and processor distribution. The results will then be automatically stored in the master processor, which has the index 0. The parallel BOR/FDTD code can handle very large problems and we have used it to simulate a reflector antenna with a radius of 1,000 wavelengths and we could handle larger problems if desired. Although the example we consider in this section can be handled on a single processor, we still employ multiple processors to simulate it in order to validate the parallel BOR/FDTD method described in this chapter.

10.4 REFLECTOR ANTENNA SYSTEM

The paraboloid reflector antenna we consider here is a rotationally symmetric structure. However, its feed system is only partially symmetric, since it is comprised of a four-element rectangular patch array. It is evident that because the system we are analyzing is not strictly circularly symmetric the conventional BOR/FDTD algorithm cannot be applied to simulate this type of problem without appropriate modifications. In this section, we will combine the BOR/FDTD with the reciprocity principle [4, 5] to simulate the partially symmetric problem.

10.4.1 Reflector Antenna Simulation

We assume that the reflector antenna, shown in Figure 10.6(a), is operating at a frequency of 45 GHz, its diameter is 160.5 mm (24.07λ), and its focal length is 90.5 mm (13.58λ). The computational domain is discretized into 500 × 340 nonuniform cells with the cell size equal to one-twentieth of the wavelength in both the ρ- and z-directions. The near-field distribution in front of reflector is shown in Figure 10.6(b) for a normally incident plane wave excitation and with the feed removed. In order to model the curved PEC surface of the reflector antenna, we employ the conformal technique described in Chapter 3. The total field distribution across the focal plane obtained by using the staircasing approximation as well as the conformal technique are shown in Figure 10.7. We observe from the above figure that the use of the conformal technique leads to a significant improvement over the staircasing approximation.

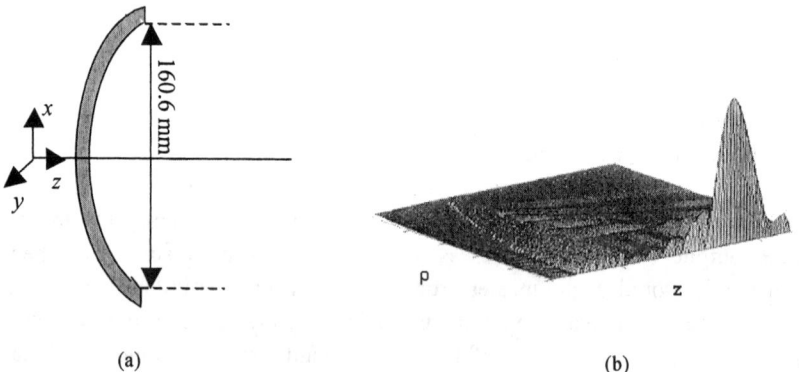

Figure 10.6 Reflector antenna and field distribution in front of reflector. (a) Dimension of reflector antenna. (b) Field distribution in front of the reflector antenna.

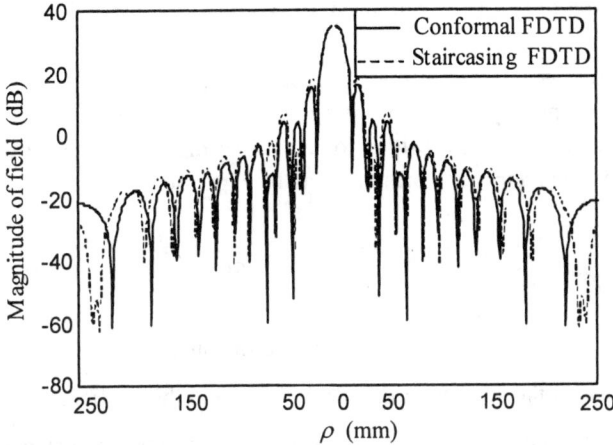

Figure 10.7 Comparison between the staircasing approximation and the conformal technique.

Next, we simulate the reflector antenna illustrated by an obliquely incident plane wave. For the sake of simplicity, we only consider a small incident angle of 3°, so that the convergent solution requires the first seven terms given in (9.56) and (9.57). The total field distribution across the focal point is plotted in Figure 10.8, together with the corresponding distribution for the normally incident case for comparison.

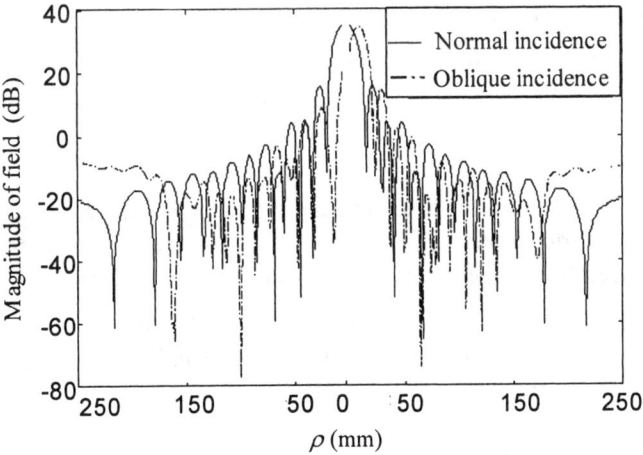

Figure 10.8 Total field distribution for an oblique incident plane wave ($\theta = 3°$).

10.4.2 Combination with Reciprocity Principle

In this section, we combine the reciprocity principle and the BOR/FDTD to simulate an antenna system comprising of the paraboloid reflector with a 2×2 rectangular patch array feed residing on a circular dielectric slab backed by a PEC ground plane. The reciprocity principle can be expressed as [6]:

$$\iint_{S_1} \vec{J}_1 \cdot \vec{E}_2 \, ds = \iiint_{V_2} \vec{J}_2 \cdot \vec{E}_1 \, dv \qquad (10.1)$$

where S_1 is an area of the patch antenna array in the feed structure, $\vec{J}_1$ is the current density on the surface of the patch array, and the far field $\vec{E}_1$ (unknown) is generated by the current density $\vec{J}_1$. While V_2 is a volume that encloses a test dipole, $\vec{J}_2$ is the current density at the location of the test dipole, and the near field $\vec{E}_2$ in position of the patch array is generated by $\vec{J}_2$. Note that we have replaced the volume integral at the left-hand side of (10.1) with a surface integral over S_1 because $\vec{J}_1$ is a surface current density.

In the current example, we first use the 3-D FDTD method to calculate the electric current density $\vec{J}_1$ on the surface of the patch array for a given excitation. Next, we use the parallel BOR/FDTD method to calculate the electric field $\vec{E}_2$ at the location of the patch array generated by a plane wave corresponding $\vec{J}_2$ (the patch array are removed from the computational domain), then calculate the far field $\vec{E}_1$ through the reciprocity principle. The step-by-step procedure is given here:

Step 1: Use a 3-D FDTD code to simulate the patch antenna array, and calculate the electric current density J_x and J_y. Use the electric current density to replace the patch elements by applying the equivalent principle.

Step 2: Place the feed system and reflector antenna in the computational domain. The structure, including the circular ground PEC and the dielectric layer, is now rotationally symmetric since the patch elements have been removed. Thus, we can use the parallel BOR/FDTD method to simulate the rotationally symmetric structure, and calculate the electric field distribution at the point where the patches were located.

To compute $\vec{E}_1$, the far field generated by the microstrip array feed radiating in the presence of the reflector, one strategy is to use (10.1). Thus, we need to calculate $\vec{E}_2$ for different incident angles of the plane waves impinging on the reflector. This, in turn, requires us to run the BOR/FDTD code repeatedly for each angle of incidence, although the current density distribution $\vec{J}_1$ obtained in Step 1 remains unchanged. We also point out that for a given incidence angle we need to run the BOR/FDTD code separately for each mode, and then sum up their contributions to obtain the total far field.

In most practical problems, we are also interested in determining the axial ratio of the radiated field, derivable from the knowledge of the co- and cross-polarization levels. Usually, we use E_{nm} to denote the n-polarized field generated by the m-polarized current source. Using the notation we interpret E_{xx} as the copolarized component radiated by J_x, while E_{yx} is the cross-polarized component; similar definitions apply to J_y as well.

Step 3: For electric circularly polarized fields, we define the co- and cross-polarizations as follows:

$$E_x^{\text{co-pol}} = E_{xx} - jE_{xy} \tag{10.2}$$

$$E_y^{\text{co-pol}} = E_{yx} - jE_{yy} \tag{10.3}$$

$$E_x^{\text{cross-pol}} = E_{xx} + jE_{xy} \tag{10.4}$$

$$E_y^{\text{cross-pol}} = E_{yx} + jE_{yy} \tag{10.5}$$

Step 4: Utilize the reciprocity principle to achieve the far field generated by the current distribution on the patch surface:

$$E_{\text{far, co-pol}} = \int_S \left(E_x^{\text{co-pol}} \cdot J_x + E_y^{\text{co-pol}} \cdot J_y \right) ds \tag{10.6}$$

$$E_{\text{far, cross-pol}} = \int_S \left(E_x^{\text{cross-pol}} \cdot J_x + E_y^{\text{cross-pol}} \cdot J_y \right) ds \tag{10.7}$$

where S is the surface area of the patch elements.

10.4.3 Simulation of the Reflector Antenna System

The method outlined in the last section was used to simulate the geometry shown in Figure 10.9. The expanded view of the patch antenna feed is depicted in Figure 10.10.

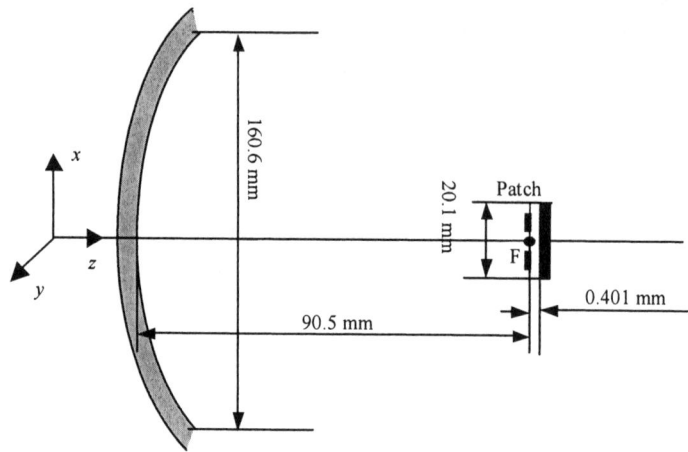

Figure 10.9 Configuration of the reflector antenna system.

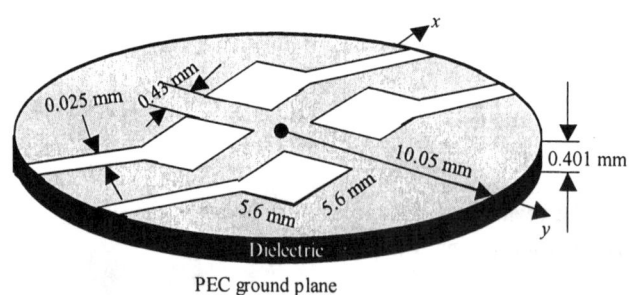

Figure 10.10 Structure of the patch antenna feed.

The normalized far-field patterns are plotted in Figure 10.11 for both co- and cross-polarizations, and are also compared with the measured results.

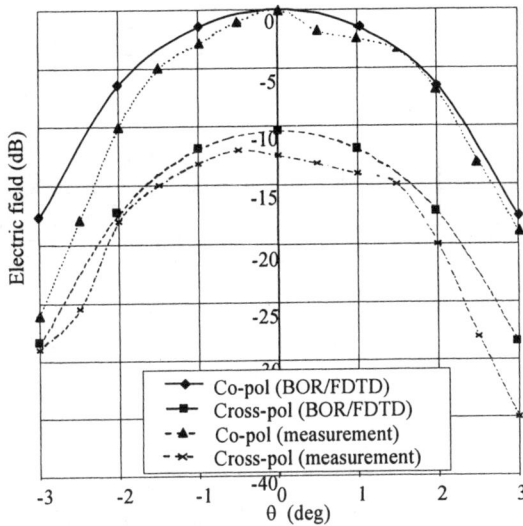

Figure 10.11 Far-field pattern of the reflector antenna system.

As a final example, we use the parallel BOR/FDTD method in conjunction with the reciprocity principle to simulate the reflector antenna system, including a radome whose geometry is shown in Figure 10.12.

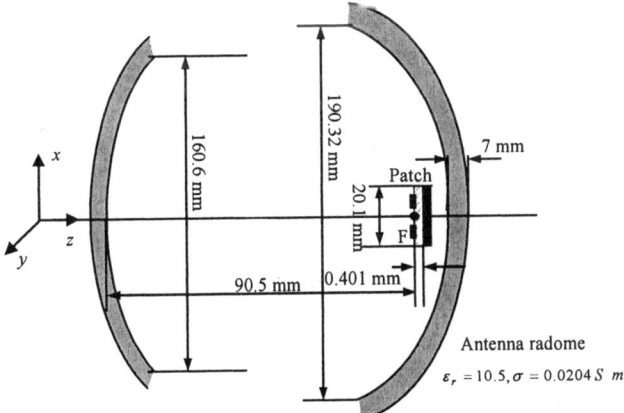

Figure 10.12 Configuration of a reflector antenna with radome.

The diameter of the antenna radome is 190.32 mm (28.55λ). The procedure, described previously, is followed once again to obtain the normalized far fields plotted in Figure 10.13, for both the co- and cross-polarizations.

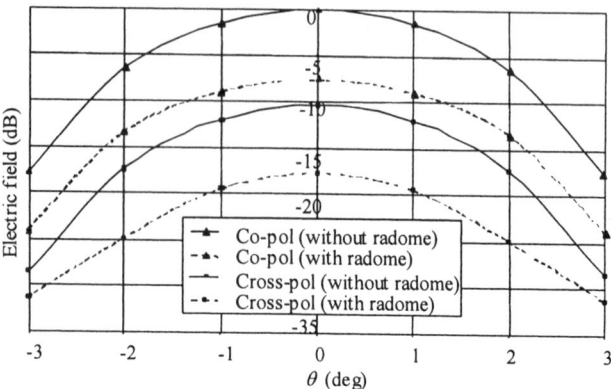

Figure 10.13 Far-field pattern for the reflector antenna system.

Although the example shown above only falls in the category of a moderately large problem, the BOR/FDTD code can handle problem geometries that are much larger, provided they possess an azimuthal symmetry at least partially, and are amenable to treatment using a combination of the BOR/FDTD and the reciprocity principle.

REFERENCES

[1] W. Gropp, E. Lusk, and A. Skjellum, *Using MPI: Portable Parallel Programming with the Message-Passing Interface*, 2nd ed., MIT Press, Cambridge, MA, 1999.

[2] W. Yu, et al., "A Robust Parallelized Conformal Finite Difference Time Domain Field Solver Package Using the MPI Library," *IEEE Antennas and Propagation Magazine*, Vol. 47, No. 3, 2005, pp. 39-59.

[3] P. Pacheco, *Parallel Programming with MPI*, Morgan Kaufmann Publishers, San Francisco, CA, 1997.

[4] W. Yu, D. Arakaki, and R. Mittra, "On the Solution of a Class of Large Body Problems with Full or Partial Circular Symmetry by Using the Finite Difference Time Domain (FDTD) Method," *IEEE Transactions on Antennas and Propagation*, Vol. 48, No. 12, 2000, pp. 1810-1817.

[5] D. Arakaki, W. Yu, and R. Mittra, "On the Solution of a Class of Large Body Problems with Partial Circular Symmetry (Multiple Asymmetries) by Using a Hybrid Dimensional Finite Difference Time Domain (FDTD) Method," *IEEE Transactions on Antennas and Propagation*, Vol. 49, No. 3, March 2001, pp. 354-360.

[6] R. Harrington, *Time-Harmonic Electromagnetic Fields*, IEEE Press, Piscataway, NJ, 2001.

Appendix A

Introduction to Basic MPI Functions

The Message-Passing Interface (MPI) library [1–3] is a standard for message passing as opposed to being a programming language. The MPI library has been widely used in a variety of parallel processing systems as an international standard, and the latest version of this MPI library, namely MPI 2.0, has recently been released. MPI 2.0 offers considerably more functions than did the MPI 1.0, though many of them are not often used in the FDTD codes.

One popular implementation of the MPI library is the MPICH [4], which was developed by the University of Chicago and Argonne National Laboratory. The MPICH can be freely downloaded online from the official Web site. The MPICH can be used on Windows and Linux, as well as Unix systems.

The MPI library includes about 210 functions. However, we have used only a dozen or so in the FDTD development. In this appendix, we will introduce some of the functions that are commonly used in the parallel FDTD code development.

A.1 SELECTED MPICH FUNCTIONS IN FORTRAN

Routine	MPI_Abort	
Format	MPI_Abort(comm, errorcode, ierr)	
Parameter	integer	comm, errorcode, ierr
Function	Terminates MPI execution environment	
Explanation	Comm	communicator of tasks to abort
	errorcode	error code to return to invoking environment
	Ierr	Routine calling success or not
Routine	MPI_Allgather	
Format	MPI_Allgather(sendbuf, sendcount, sendtype, recvbuf, recvcount, recvtype, comm, ierr)	

Parameter	<type>	sendbuf(*), recvbuf(*)
	integer	sendcount, sendtype, recvcount, recvtype, comm, ierr
Function	Gathers data from all tasks and distributes it to all	
Explanation	Sendbuf	starting address of send buffer
	sendcount	number of elements in send buffer
	sendtype	data type of send buffer elements
	Recvbuf	starting address of receive buffer
	recvcount	number of elements received from any process
	Recvtype	data type of receive buffer elements
	Comm	communicator
Routine	MPI_Allgatherv	
Format	MPI_Allgatherv (sendbuf, sendcount, sendtype, recvbuf, recvcounts, displs, recvtype, comm, ierr)	
Parameter	<type>	sendbuf(*), recvbuf(*)
	integer	sendcount, sendtype, recvcount(*), displs(*), recvtype, comm, ierr
Function	Gathers data from all tasks and delivers it to all	
Explanation	Sendbuf	starting address of send buffer
	sendcount	number of elements in send buffer
	sendtype	data type of send buffer elements
	Recvbuf	starting address of receive buffer
	recvcount	number of elements received from any process
	Displs	integer array (of length group size)
	Recvtype	data type of receive buffer elements
	Comm	communicator
Routine	MPI_Allreduce	
Format	MPI_Allreduce(sendbuf, recvbuf, count, datatype, op, comm, ierr)	
Parameter	<type>	sendbuf(*), recvbuf(*)
	integer	count, datatype, op, comm, ierr
Function	Combines values from all processes and distributes the result back to all processes	
Explanation	sendbuf	starting address of send buffer
	Recvbuf	starting address of receive buffer
	Count	number of elements in send buffer
	Datatype	data type of elements of send buffer
	Op	operation
Routine	MPI_Barrier	

Format	MPI_Barrier (comm, ierr)
Parameter	Integer comm, ierr
Function	Blocks until all processes have reached this routine

Routine	MPI_Bcast	
Format	MPI_Bcast(buffer, count, datatype, root, comm, ierr)	
Parameter	\<type\> buffer(*)	
	integer count, datatype, root, comm, ierr	
Function	Broadcasts a message from the process with rank "root" to all other processes of the group	
Explanation	Buffer	starting address of buffer
	Count	number of entries in buffer
	datatype	data type of buffer
	Root	rank of broadcast root

Illustration
count = 1
root = 1
MPI_Bacast (buffer, count, MPI_INTEGER, root, MPI_COMM_WORLD, ierr)

Task 0   Task 1   Task 2   Task 3

| | 7 | | | Msg (before) |

| 7 | 7 | 7 | 7 | Msg (after) |

Routine	MPI_Comm_rank
Format	MPI_Comm_rank(comm, rank, ierr)
Parameter	Integer comm, rank, ierr
Function	Determines the rank of the calling process in the communicator
Explanation	Rank rank of the calling process in group of comm

Routine	MPI_Comm_size
Format	MPI_Comm_size(comm, size, ierr)
Parameter	Integer comm, size, ierr
Function	Determines the size of the group associated with a communictor
Explanation	Size number of processes in the group of comm

Routine	MPI_Finalize	
Format	MPI_Finalize(ierr)	
Parameter	Integer ierr	
Function	Terminates MPI execution environment	

Routine	MPI_Gather	
Format	MPI_Gather(sendbuf, sendcount, sendtype, recvbuf, recvcount, recvtype, dest, comm, ierr)	
Parameter	<type>	sendbuf(*), recvbuf(*)
	integer	sendcount, sendtype, recvcount, recvtype, dest, comm, ierr
Function	Gathers together values from a group of processes	
Explanation	sendbuf	starting address of send buffer
	sendcount	number of elements in send buffer
	sendtype	data type of send buffer elements
	recvbuf	address of receive buffer
	recvcount	number of elements for any single receive
	recvtype	data type of recv buffer elements
	Dest	rank of receiving process

Illustration
sendcount = 1
recvcount = 1
dest = 1
MPI_Gather (sendbuf, sendcount, MPI_INTEGER, recvbuf, recvcount, MPI_INTEGER, dest, MPI_COMM_WORLD, ierr)

task0	task1	task2	task3	
1	2	3	4	Sendbuf (before)

	1			
	2			Recvbuf (before)
	3			
	4			

Introduction to Basic MPI Functions

Routine	MPI_Gatherv
Format	MPI_Gatherv (sendbuf, sendcount, sendtype, recvbuf, recvcounts, displs, recvtype, dest, comm, ierr)
Parameter	<type> sendbuf(*), recvbuf(*)
	integer sendcount, sendtype, recvcount(*), displs(*), recvtype, dest, comm, ierr
Function	Gathers into specified locations from all processes in a group
Explanation	Sendbuf starting address of send buffer
	sendcount number of elements in send buffer
	Sendtype data type of send buffer elements
	Recvbuf address of receive buffer
	recvcount integer array (of length group size) containing the number of elements that are received from each process
	Displs integer array (of length group size). Entry i specifies the displacement relative to recvbuf at which to place the incoming data from process i (significant only at root)
	Recvtype data type of recv buffer elements
	Dest rank of receiving process

Routine	MPI_Init
Format	MPI_Init(ierr)
Parameter	integer Ierr
Function	Initializes the MPI execution environment

Routine	MPI_Irecv
Format	MPI_Irecv(buffer, count, datatype, source, tag, comm, request, ierr)
Parameter	<type> buffer(*)
	integer count, datatype, source, tag, comm, request, ierr
Function	Begins a nonblocking receive
Explanation	buffer address of receive buffer
	count number of elements in receive buffer
	datatype datatype of each receive buffer element
	source rank of source
	tag message tag
	request communication request

Routine	MPI_Isend	
Format	MPI_Isend (buffer, count, datatype, dest, tag, comm, request, ierr)	
Parameter	<type>	buffer(*)
	integer	count, datatype, dest, tag, comm, request, ierr
Function Explanation	Begins a nonblocking send	
	buffer	starting address of send buffer
	count	number of elements in send buffer
	datatype	datatype of each send buffer element
	dest	rank of destination
	tag	message tag
	request	communication request
Routine	MPI_Reduce	
Format	MPI_Reduce (sendbuf, recvbuf, count, datatype, op, dest, comm, ierr)	
Parameter	<type>	sendbuf(*), recvbuf(*)
	integer	count, datatype, op, dest, comm, ierr
Function Explanation	Reduces values on all processes to a single value	
	sendbuf	starting address of send buffer
	recvbuf	address of receive buffer
	count	number of elements in send buffer
	datatype	data type of elements of send buffer
	dest	rank of destination process
	op	reduce operation

Illustration
count = 1
dest = 1
MPI_Reduce (sendbuf, recvbuf, count, MPI_INTEGER, MPI_SUM, dest, MPI_COMM_WORLD, ierr)

task0	task1	task2	task3	
1	2	3	4	Sendbuf (before)
	10			Recvbuf (before)

Routine   MPI_Type_commit

Introduction to Basic MPI Functions

Format	MPI_Type_commit (datatype, ierr)
Parameter	integer datatype, ierr
Function	Commits the datatype
Explanation	datatype Datatype

Routine	MPI_Type_free
Format	MPI_Type_free (datatype, ierr)
Parameter	integer datatype, ierr
Function	Frees the datatype
Explanation	datatype datatype that is freed

Routine	MPI_Type_struct
Format	MPI_Type_struct(count, blocklens, indices, old_types, newtype, ierr)
Parameter	integer count, blocklens(*), indices(*), old_types(*), newtype, ierr
Function	Creates a struct datatype
Explanation	count number of blocks
	blocklens number of elements in each block
	indices byte displacement of each block
	old_type type of elements in each block
	newtype new datatype

Routine	MPI_Type_vector
Format	MPI_Type_vector(count, blocklen, stride, old_type, newtype, ierr)
Parameter	integer count, blocklen, stride, old_type, newtype, ierr
Function	Creates a vector (strided) datatype
Explanation	count number of blocks
	blocklen number of elements in each block
	stride number of elements between start of each block
	old_type type of elements in each block
	newtype new datatype

Routine	MPI_Wait
Format	MPI_Wait(request, status, ierr)
Parameter	integer request, status, ierr
Function	Waits for an MPI send or receive to complete
Explanation	request communication request
	status status object

A.2 SELECTED MPICH FUNCTIONS IN C/C++

Routine	MPI_Abort	
Format	int MPI_Abort(MPI_Comm comm, int errorcode)	
Function	Terminates MPI execution environment	
Explanation	comm	communicator of tasks to abort
	errorcode	error code to return to invoking environment

Routine	MPI_Allgather	
Format	int MPI_Allgather (void *sendbuf, int sendcount, MPI_Datatype sendtype, void *recvbuf, int recvcount, MPI_Datatype recvtype, MPI_Comm comm)	
Function	Gathers data from all tasks and distributes it to all	
Explanation	sendbuf	starting address of send buffer
	sendcount	number of elements in send buffer
	sendtype	data type of send buffer elements
	recvbuf	starting address of receive buffer
	recvcount	number of elements received from any process
	recvtype	data type of receive buffer elements
	comm	Communicator

Routine	MPI_Allgatherv	
Format	int MPI_Allgatherv (void *sendbuf, int sendcount, MPI_Datatype sendtype, void *recvbuf, int *recvcounts, int *displs, MPI_Datatype recvtype, MPI_Comm comm)	
Function	Gathers data from all tasks and delivers it to all	
Explanation	sendbuf	starting address of send buffer
	sendcount	number of elements in send buffer
	sendtype	data type of send buffer elements
	recvbuf	starting address of receive buffer
	recvcount	number of elements received from any process
	displs	integer array (of length group size)
	recvtype	data type of receive buffer elements
	comm	Communicator

Routine	MPI_Allreduce
Format	int MPI_Allreduce (void *sendbuf, void *recvbuf, int count, MPI_Datatype datatype, MPI_Op op, MPI_Comm comm)

Function	Combines values from all processes and distribute the result back to all processes
Explanation	sendbuf starting address of send buffer
	recvbuf starting address of receive buffer
	count number of elements in send buffer
	datatype data type of elements of send buffer
	op Operation

Routine	MPI_Barrier
Format	int MPI_Barrier (MPI_Comm comm)
Function	Blocks until all process have reached this routine

Routine	MPI_Bcast
Format	int MPI_Bcast (void *buffer, int count, MPI_Datatype datatype, int source, MPI_Comm comm)
Function	Broadcasts a message from the process with rank " root " to all other processes of the group
Explanation	buffer starting address of buffer
	count number of entries in buffer
	datatype data type of buffer
	source rank of broadcast root

Illustration
count = 1
source = 1
MPI_Bacast (buffer, count, MPI_INT, source, MPI_COMM_WORLD)

Task 0	Task 1	Task 2	Task 3	
	7			Msg (before)
7	7	7	7	Msg (after)

Routine	MPI_Comm_rank
Format	int MPI_Comm_rank (MPI_Comm comm, int *rank)
Function	Determines the rank of the calling process in the communicator
Explanation	rank rank of the calling process in group of comm

Routine	MPI_Comm_size
Format	int MPI_Comm_size (MPI_Comm comm, int *size)
Function	Determines the size of the group associated with a communicator
Explanation	size number of processes in the group of comm

Routine	MPI_Finalize
Format	int MPI_Finalize()
Function	Terminates MPI execution environment

Routine	MPI_Gather
Format	int MPI_Gather (void *sendbuf, int sendcnt, MPI_Datatype sendtype, void *recvbuf, int recvcount, MPI_Datatype recvtype, int dest, MPI_Comm comm)
Function	Gathers together values from a group of processes
Explanation	sendbuf starting address of send buffer sendcount number of elements in send buffer sendtype data type of send buffer elements recvbuf address of receive buffer recvcount number of elements for any single receive recvtype data type of recv buffer elements dest rank of receiving process

Illustration
sendcount = 1
recvcount = 1
dest = 1
MPI_Gather(sendbuf, sendcount, MPI_INT, recvbuf, recvcount, MPI_INT, dest, MPI_COMM_WORLD)

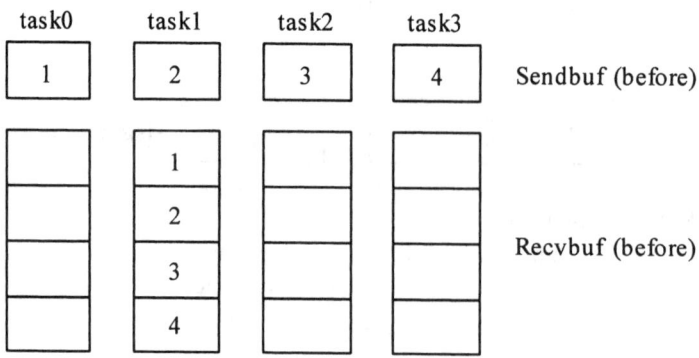

Routine	MPI_Gatherv
Format	int MPI_Gatherv (void *sendbuf, int sendcnt, MPI_Datatype sendtype, void *recvbuf, int *recvcnts, int *displs, MPI_Datatype recvtype, int dest, MPI_Comm comm)
Function	Gathers into specified locations from all processes in a group
Explanation	sendbuf starting address of send buffer
	sendcnt number of elements in send buffer
	sendtype data type of send buffer elements
	recvbuf address of receive buffer
	recvcnts integer array (of length group size) containing the number of elements that are received from each process
	displs integer array (of length group size). Entry i specifies the displacement relative to recvbuf at which to place the incoming data from process i (significant only at root)
	recvtype data type of recv buffer elements
	dest rank of receiving process
Routine	MPI_Init
Format	int MPI_Init(int *argc, char ***argv)
Explanation	These two parameters are the parameter of main Routine
Function	Initializes the MPI execution environment
Routine	MPI_Irecv
Format	int MPI_Irecv(void *buffer, int count, MPI_Datatype datatype, int source, int tag, MPI_Comm comm, MPI_Request *request)
Function	Begins a nonblocking receive
Explanation	buffer address of receive buffer
	count number of elements in receive buffer
	datatype datatype of each receive buffer element
	source rank of source
	tag message tag
	request communication request
Routine	MPI_Isend
Format	int MPI_Isend(void * buffer, int count, MPI_Datatype datatype, int dest, int tag, MPI_Comm comm, MPI_Request *request)
Function	Begins a nonblocking send

Explanation	buffer	starting address of send buffer
	count	number of elements in send buffer
	datatype	datatype of each send buffer element
	dest	rank of destination
	tag	message tag
	request	communication request
Routine	MPI_Reduce	
Format	int MPI_Reduce (void *sendbuf, void *recvbuf, int count, MPI_Datatype datatype, MPI_Op op, int dest, MPI_Comm comm)	
Function	Reduces values on all processes to a single value	
Explanation	sendbuf	starting address of send buffer
	recvbuf	address of receive buffer
	count	number of elements in send buffer
	datatype	data type of elements of send buffer
	dest	rank of destination process
	op	reduce operation

Illustration
count = 1
dest = 1
MPI_Reduce (sendbuf, recvbuf, count, MPI_INTEGER, MPI_SUM, dest, MPI_COMM_WORLD, ierr)

task0   task1   task2   task3

| 1 | | 2 | | 3 | | 4 | Sendbuf (before)

| | | 10| | | | | Recvbuf (before)

Routine	MPI_Type_commit	
Format	int MPI_Type_commit (MPI_Datatype *datatype)	
Function	Commits the datatype	
Explanation	datatype	Datatype

Routine	MPI_Type_free	
Format	int MPI_Type_free (MPI_Datatype *datatype)	
Function	Frees the datatype	
Explanation	datatype	datatype that is freed

Routine	MPI_Type_struct
Format	int MPI_Type_struct(int count, int blocklens[], MPI_Aint indices[], MPI_Datatype old_types[], MPI_Datatype *newtype)
Function	Creates a struct datatype
Explanation	count — number of blocks blocklens — number of elements in each block indices — byte displacement of each block old_type — type of elements in each block newtype — new datatype

Routine	MPI_Type_vector
Format	int MPI_Type_vector(int count, int blocklen, int stride, MPI_Datatype old_type, MPI_Datatype *newtype)
Function	Creates a vector (strided) datatype
Explanation	count — number of blocks blocklen — number of elements in each block stride — number of elements between start of each block old_type — type of elements in each block newtype — new datatype

Routine	MPI_Wait
Format	int MPI_Wait (MPI_Request *request, MPI_Status *status)
Function	Waits for an MPI send or receive to complete
Explanation	request — communication request status — status object

A.3 MPI DATA TYPE

C/C++ Data Type		Fortran Data Type	
MPI_CHAR	signed char	MPI_CHARACTER	character(1)
MPI_SHORT	signed short int		
MPI_INT	signed int	MPI_INTEGER	integer
MPI_LONG	signed long int		
MPI_UNSIGNED_CHAR	unsigned char		
MPI_UNSIGNED_SHORT	unsigned short int		
MPI_UNSIGNED	unsigned int		

MPI_UNSIGNED_LONG	unsigned long int		
MPI_FLOAT	float	MPI_REAL	Real
MPI_DOUBLE	double	MPI_DOUBLE_PRECISION	double precision
MPI_LONG_DOUBLE	long double		
		MPI_COMPLEX	complex
		MPI_DOUBLE_COMPLEX	double complex
		MPI_LOGICAL	logical
MPI_BYTE	8 binary digits	MPI_BYTE	8 binary digits
MPI_PACKED	data packed or unpacked with MPI_Pack() /MPI_Unpack	MPI_PACKED	data packed or unpacked with MPI_Pack() /MPI_Unpack

A.4 MPI OPERATOR

MPI Operator		C/C++ Data Type	Fortran Data Type
MPI_MAX	maximum	int, float	integer, real, complex
MPI_MIN	minimum	int, float	integer, real, complex
MPI_SUM	sum	int, float	integer, real, complex
MPI_PROD	product	int, float	integer, real, complex
MPI_LAND	logical AND	int	logical
MPI_BAND	bit-wise AND	int, MPI_BYTE	int, MPI_BYTE
MPI_LOR	logical OR	int	logical
MPI_BOR	bit-wise OR	int, MPI_BYTE	int, MPI_BYTE
MPI_LXOR	logical XOR	int	logical
MPI_BXOR	bit-wise XOR	int, MPI_BYTE	int, MPI_BYTE
MPI_MAXLOC	max value and location	float, double and long double	real, complex, double precision
MPI_MINLOC	min value and location	float, double and long double	real, complex, double precision

A.5 MPICH SETUP

As mentioned earlier, the MPICH is one of the MPI implementations and can freely be downloaded from its official Web site http://www.unix.mcs.anl.gov/mpi/mpich/. The MPICH runs on different platforms such as Windows, Linux, and UNIX.

A.5.1 MPICH Installation and Configuration for Windows

MPICH Installation:
(1) Logon to the machine using an account that has the administrator privilege. The user needs to apply for this permit or ask help of the administrator.
(2) Execute "mpich.nt.1.2.5.exe," and select all the default options.
(3) Repeat steps 1 and 2 on each machine in the cluster.

MPICH Configuration:
(1) Add the bin folder to the Windows system path. The default folder is C:\Program Files\MPICH\mpd\bin
(2) The user must register by running the MPIRegister.exe, which is located in C:\Program Files\MPICH\mpd\bin. The userID and password must be consistent with those for Windows. The password cannot be chosen to be the blank.

A.5.2 MPICH Installation and Configuration for Linux

Package uncompress:
Copy the downloaded MPICH package to a temporary folder, and then uncompress it using the following command:

tar zxvof mpich.tar.gz

If the z option is invalid in the command above, there is an alternative way for the same purpose:

gunzip -c mpich.tar.gz | tar xovf –

MPICH installation and configuration:
In Linux and Unix, we need to configure and compile the MPICH before using it. The configuration operation will analyze the system information, determine the correct options, and create a makefile. Follow the steps below to install and configure the MPICH:
(1) Select an MPICH installation folder. By the way, it would be better for the user to use the version serial number in the MPICH folder name, for instance, /usr/local/mpich-1.2.5, so that he/she can install the different version of the MPICH.

(2) Use the configuration command to configure the MPICH. The default login shell is the Remote Shell, but the default status of this shell is closed in most Linux systems. Consequently, we usually use the Secure Shell instead of the Remote Shell. We note here that the ssh is actually more secure than the above two options. We can specify Fortran 77, Fortran 90, or C/C++ compiler during the process of MPICH configuration. The desired user can refer to the MPICH manual for more details. The configuration command may read as:

./configure –fc=ifc –f90=ifc –rsh=ssh –prefix=/usr/local/mpich-1.2.5 | tee c.log

(3) MPICH compilation and linking can be done using the following command:

make | tee make.log

(4) MPICH can be installed through the following command:

make install

Configuration of the MPICH after installation:

(1) Add the MPICH folder to the system path. First, the user needs to edit the file /etc/profile, and then adds the MPICH folder into the path of the profile file. If the MPICH is just for the user himself/herself, the user can edit the file ~/.profile, and then add the MPICH folder into the path of the profile file.

(2) Configure ssh for each user using the following commands:

ssh-keygen -t rsa
cp ~/.ssh/rsa.pub ~/.ssh/authorized_keys
chmod 755 ~/.ssh
ssh-agent $SHELL
ssh-add

REFERENCES

[1] W. Gropp, et al., *Using MPI: Portable Parallel Programming with the Message-Passing Interface*, 2nd ed., MIT Press, Cambridge, MA, 1999.

[2] P. Pacheco, *Parallel Programming with MPI*, Morgan Kaufmann Publishers, San Francisco, CA, 1997.

[3] M. Snir, et al., *MPI: The Complete Reference*, Vols. I and II, MIT Press, Cambridge, MA, 1998.

[4] MPICH (MPI Chameleon), Argonne National Laboratory: http://www.mcs.anl.gov/.

Appendix B

PC Cluster-Building Techniques

High performance computational PC clusters are configured by connecting personal computers via a network [1–3]. In this appendix, we present the detailed procedure that is used to set up a cluster. In addition, we also briefly introduce the basic requirements for running a parallel FDTD code on an existing Beowulf PC cluster.

B.1 SETUP OF A PC CLUSTER FOR THE LINUX SYSTEM

To build a PC cluster, the first thing we must consider is the node hardware. A node is a stand-alone computing system with a complete hardware and operating system that can run the user's application codes and communicate with other nodes through network. A cluster, which contains the identical nodes including CPU, RAM, and all other components except hard disk size, is called an isomorphic cluster. Otherwise, the cluster is referred to as heteromorphic. The advantage of an isomorphism cluster is that it does not require the user to be familiar with the structure of the cluster for job balancing.

An important factor that affects the cluster efficiency is the network device, which includes the network interface controller (NIC), HUB, and the switch. Traditionally, the network device is considered as a major bottleneck in the parallel computation schemes. However, the network devices have been fast developing in recent years — the NIC and Myricom switches can reach a speed up to 10 Gbps, for instance. Since the high-performance network device is very expensive, the user should select it with care.

The NIC is a device installed in the node that accepts the message from the host node, and then delivers the data to the physical network media. The switch and HUB are the devices that accept packages from the nodes, but the HUB sends package to all the nodes. However, the switch only sends packages to the target nodes. All the nodes connected to the HUB share its bandwidth, while the nodes connected to a switch have the same bandwidth as that of the switch. We usually use a fast switch in the cluster instead of the HUB. A typical PC cluster is shown in Figure B.1.

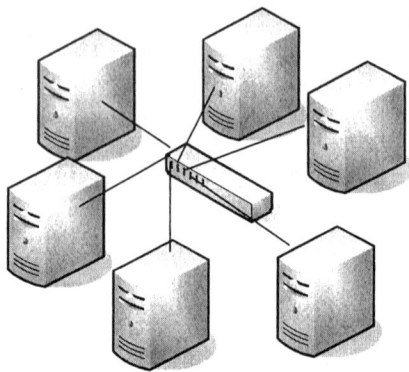

Figure B.1 Illustration of the simple cluster.

B.1.1 PC Cluster System Description

A typical PC cluster consists of the following software and hardware:

Login node: Dell Precision 340, CPU Pentium4 3.0 GHz, 2 GB RAM, 160 GB hard drive, one 1 GB Ethernet NIC connecting the inner node of the cluster, one 100 MB Ethernet NIC connecting outer LAN.
Compute node: Dell Precision 340, CPU Pentium4 3.0 GHz, 2 GB RAM, 1 GB Ethernet NIC, the hard drive of compute node doesn't need to be very large, 20 GB is large enough to install the OS.
Switch: Dell PowerConnect 2508, 1 GB, 8-port network switch
Cable: Category 5 UTP
Linux OS: Redhat Linux Ver9.0 or Redhat AS

B.1.2 Linux Installation

Before we install the Linux operation system, we need to connect all the nodes to the switch, and the 100 MB NIC of the login node to other LAN.

If the CD-ROM is not the default boot device, we should set it to be the default from the BIOS. After booting, we will see a boot prompt and instructions on the screen. The default installation mode is the graphical mode, and can be changed to the text mode. We can select the mouse, the keyboard, the monitor, and the graphics card to configure the system.

During installation of the Linux system, we need to partition the hard disk. We can configure the login node that contains the system software, and a large disk for the home directories. Home directories should be created in a separate partition that is easy to backup, upgrade, and maintain.

A swap partition should be created even when we have a large memory. The swap space should be twice the physical memory. However, it is not larger than 2 GB for the 32-bit Intel processors. A typical partition of the login node is shown in Table B.1.

Table B.1
A Typical Partition of Login Node

	Mount Point	File System Type	Size
sda1	/boot	ext3	102 MB
sda2	/	ext3	20 GB
sda3		swap	2 GB
sda4	/home	ext3	140 GB

For the computing node, we can use automatic partition if the hard drive is not too large, otherwise, the partition should be created in a manner shown in Table B.2.

Table B.2
A Typical Partition of Computing Node

	Mount Point	File System Type	Size
Sda1	/boot	ext3	102 MB
Sda2	/	ext3	20 GB
Sda3		swap	2 GB
Sda4	/scratch	ext3	140 GB

A scratch partition is a space that is used to store the user's temporary data generated during the simulation. The home space for each user usually has a limited size; however, the scratch space is sufficient for most projects. Generally speaking, the user manages home space and the administrator controls the scratch space, which is cleaned up periodically.

After partitioning the hard disk we need to format the partition. If we reinstall the Linux system, we should not format the home partition, otherwise, all the user's data in the home partition will be lost.

In addition to the system files, we must install some software that we need, for instance programming languages such as C/C++ and/or Fortran compilers, network packages such as the Web server, the DNS server, and the mail server.

The TCP/IP is the standard network communication protocol. The Internet protocol packet is specified by a 32-bit long IP address that uniquely identifies the host machine. So each node in the cluster must be assigned an IP address. An IP address includes a network address and the host address. The division between two parts is specified by the netmask. A typical netmask is 255.255.255.0, which specifies 24 bits of network addresses and eight bits of host addresses. Three IP address ranges have been reserved for the private networks:

10.0.0.0 – 10.255.255.255
172.16.0.0 – 172.31.255.255
192.168.0.0 – 192.168.255.255

These address ranges are permanently unassigned and will not be forwarded by Internet backbone routers, otherwise, they will conflict with publicly accessible IP address.

We use 192.168.1.0–192.168.1.255 IP address range for our cluster node IP assignment. We should note that 192.168.1.255 is the network broadcast address, packets sent to this address will be forwarded to every host in the network, and 192.168.1.0 is also a reserved IP address. We can start with 192.168.1.1, which is assigned to the 1 GB NIC of login node, assign 192.168.1.2 to the first computing node, and so on.

Generally speaking, the login node has two Ethernet NICs: one is used for inner cluster communication, and another one is for outside cluster communication. We need to assign the second NIC with an IP address that belongs to LAN so that the cluster can communicate with other networks.

We should specify a hostname to all the nodes and can select any name for the cluster nodes.

The new version of Linux has two boot loaders, the LILO and the GRUB, and we can select either of them. For the selected boot loader, we can select either the text or graphical start mode. In order to reduce the computer resource requirement, it is best to use the text mode on the computing node.

B.1.3 Linux System Configuration

After installing the Linux system, it is necessary to carry out some post-configurations for the parallel running environment. The most important files include the passwd, group, inittab, fstab, and hosts in the folder "/etc". The passwd file stores the user account, password, home directory, and shell information. The group file contains a list of user groups of the system. The inittab file is essential to properly booting a system. The fstab file defines a file system that can be mounted on the system. The host file statically lists hostname to IP address mappings and we can add the following information into this file:

192.168.1.1   b0
192.168.1.2   b1
192.168.1.2   b2
...
192.168.1.n+1   bn

B.1.3.1 NIS Configuration

The user account management is usually handled in one of the two following ways: in the first way, we create an account on the login node and then copy the files in both "/etc/passwd" and "/etc/group" to each node in the cluster. If we miss one node the system will not work properly. In the second way, we use the network information service (NIS) for the user authorization.

The NIS developed by Sun Microsystems provides a simple network lookup service system, which is used to access the database of the network and pass the messages such as password and group information to all the nodes in the cluster. By doing this, the user only manages his/her account on the login node and the NIS does all other jobs for him/her. Next, we introduce the NIS server and client configurations.

NIS server configuration: because the login node provides the NIS service, it is the NIS server. Use the following steps to configure the NIS server:

- Set up the domain server:

    domainname YourNIS_DomainName

- Modify the starting script file, /etc/sysconfig/network, and add the following line in it:

    NISDOMAIN= YourNIS_DomainName

- Create two files if they do not exist:

    touch /etc/gshadow
    touch /etc/netgroup

- Initialize the NIS server using the following command:

/usr/lib/yp/ypinit –m

Press CTRL+D to complete the NIS server configuration. The system will create a directory YourNIS_DomainName in the folder /var/yp/. It contains some image files such as passwd.byname, passwd.byuid, group.byname, and group.bygid.

- Start the NIS server using the following commands:

/etc/rc.d/init.d/ypserv start
/etc/rc.d/init.d/yppasswdd start

- Add the ypserv service to startup service using the following commands:

chkconfig --add ypserv
chkconfig --level ypserv 3 on
chkconfig --level ypserv 5 on

For a new user, we need to go to the directory "/var/yp" using the command "make" to update the database. Otherwise, the user information will not be recognized by the cluster.

NIS client configuration: all computing nodes need to install the NIS Client service. Use the following steps to configure the NIS Client:

- Edit the file "/etc/yp.conf" and add the following line in it:

domain YourNIS_DomainName

- Modify the file "/etc/sysconfig/network":

NISDOMAIN= YourNIS_DomainName

- Open the file "/etc/sysconfig/network" and then the following contents will appear:

passwd files dns
group   files dns
....

The "files" followed by the option "passwd" is determined by "/etc/passwd", and the "dns" is determined by "DNS". Because of using the NIS configuration system, the "nis" should be added in this line. The modified configuration file is as fellows:

passwd nis files dns
shadow nis files dns
group   nis files dns

- Start an NIS user computer program using the following command:

/etc/rc.d/init.d/ypbind start

- Add the "ypbind" servive to automatic starting service using the following commands:

    chkconfig --add ypbind
    chkconfig --level ypbind 3 on
    chkconfig --level ypbind 5 on

B.1.3.2 NFS Server Configuration

The network file system (NFS) provides the shared service among the nodes in the cluster. A Linux system could be a NFS server or a NFS client, which means that the NFS system can export the file system to other systems and mount the file systems exported by other systems.

The NFS sever configuration can be performed via the graphical configuration interface, in which all of the instructions are graphically displayed in the graphical window. Next, we describe the manual NFS server configuration techniques. For example, in the cluster we export the "/home" directory of the login node to share with other nodes, and the other nodes mount the export of the login node. At the login node, edit /etc/exports and add the following line:

/home   hostname(rw,sync,no_wdelay,no_root_squash)

It is worthwhile mentioning the space usage in the "/etc/exports" file. The options apply only to the hostname, if there are no spaces between the hostname, the left parentheses, and between options inside the parentheses. Otherwise, the options apply to anywhere else. The hostname may be a computer's name, or a series of machines specified with wild cards or IP address. For more details, please refer to the Redhat manual. At all the computing nodes, edit the file "/etc/fstab", and add the following line:

hostname:/home /home nfs   rsize=8192,rsize=8192,intr

All the computing nodes can have access to the "/home" directory of the login node now. If we create "/scratch" partitions at a computing node, we can export these partitions to other nodes. The procedure of partition is the same as that for "/home".

B.1.4 Developing Tools

GNU Fortran and the C/C++ compilers have been embedded in the Linux system. If the user only needs C/C++ as the developing language, the GNU C/C++

compiler is sufficient, while GNU Fortran in the Linux system is a low-level compiler that is consistent with Fortran 77 but does not include features of Fortran 90.

If an Intel processor is selected as the node in the cluster, Intel C/C++, and Fortran 90 compilers will be the best choice since Intel optimizes their compilers for Intel processors.

B.1.5 MPICH

Following the instructions described in Appendix A to download and install the MPICH, we need to edit the file "/usr/local/mpich-1.2.5/share/machines.LINUX" and add the node information to this file:

b1
b2
.
bn

The command mpirun is located in the following folder:

/usr/local/mpich-1.2.5/bin

For almost all systems, we can use the command, "mpirun –np N exefile" to run the program on N processors. The command "mpirun –help" gives us a complete list of options.

The command "mpirun" will assign a job to the corresponding nodes according to the node sequence in the file "machines.LINUX". An alternative way to run parallel code requires specifying computing nodes in the command line, for example,

mpirun –np n –machinefile nodename exefile

The "nodename" in the above line includes all the hostnames of the nodes where the job will run. The above command requires that the user be familiar with the node information in the cluster. If the "/scratch" partition is not created, the user's data will be stored in his/her home directory. Otherwise, the user can use "/scratch" partition of computing nodes to store the temporary data. The user needs to copy all project files to "/scratch" partition and run the parallel code there. After the code is completed, the user needs to copy his or her data back to his or her home directory, and delete the temporary data in the "/scratch" partition.

B.1.6 Batch Processing Systems

For a small cluster, typically only one user should use it at one time, since any conflict usage of the cluster will decrease its efficiency dramatically. In order to circumvent this problem, we need to install a batch processing system (BPS). The purpose of PBS is to schedule and initiate an execution of batch jobs, and to route these jobs between hosts. There exist many BPS software; the several commonly used BPS are listed in Table B.3.

Table B.3
Commonly Used BPS Software List

Condor	Condor http://www.cs.wisc.edu/condor/	University of Wisconsin
NQS	Network Queuing System http://www.gnqs.org/	Sun Microsystems, University of Sheffield
PBS	Portable Batch System http://pbs.mrj.com/service.html	NASA, LLNL
DQS	Distributed Queuing System http://www.genias.de/welcome.html	Florida State University
EASY-LL	Extensible Argonne Scheduling System – LoadLeveler http://www.tc.cornell.edu/Software/EASY-LL/	Cornell University, IBM
CODINE	Distributed Queuing System http://www.scri.fsu.edu/~pasko/dqs.html	Genias Software

After the BPS installation, all the parallel jobs must be submitted via the BPS, hence the system does not allow the user to run a parallel code manually. The user can compile his/her parallel code at the login node, but then must use the qsub command to submit the job. After the job is submitted, the user can use the qstat command to view the status of the job in the queue. If the user wants to kill his or her job, he or she can use the qdel command to delete the job in the queue.

B.2 PARALLEL COMPUTING SYSTEM BASED ON WINDOWS

In this section, we introduce the parallel development environment based on the Windows system. Different from the high-performance PC cluster we described in the last section, we use the following two ways to build a parallel computing system.

B.2.1 System Based on Windows

First, we need a server installed with either Windows 2000 server or Windows 2003 server operating system, to set up a domain, and then to connect all the computing nodes to this domain. In addition, we also need to install C/C++ or/and Fortran languages compiler and the MPICH in each computer in this domain.

B.2.2 System Based on a Peer Network

A peer network is rather simple compared to other systems and does not include any server. This network requires Windows 2000/XP, and all the computers in this network must be the same.

A parallel computing system based on the peer network requires that the same userID be created in each computer and all the userIDs must have the same password.

Parallel code execution in the computing system based on Windows has the same procedure as that in the Linux system. The only difference is that the user needs to create a directory in each computer and must have the same directory name in the same hard drive. The parallel codes and project files should be copied to the corresponding directory before running the code.

B.3 USING AN EXISTING BEOWULF PC CLUSTER

In this section, we briefly introduce how to run a parallel code on the existing Beowulf PC cluster, which is the most popular system in the research universities and institutes.

To run a parallel code on these systems, we assume that the system already has the necessary software, development tools, and the MPICH as well as MPI library. Otherwise, the user should send a request to the system administrator to install the necessary software.

Due to the differences from one system to another, we need to learn how to compile, link, and run the parallel code on the different systems from the available help files. Since the Beowulf PC cluster uses the BPS, we have to submit the job via the BPS. The procedure for job submitting and job status checking is similar to that described in Section B1.

The command qsub is used to submit a job to the queue. A typical PBS script file is shown below:

```
# This is a sample PBS script. It will request 4 processors on 2 nodes
# for 4 hours.
# Request 4 processors on 2 nodes
# PBS -l nodes=2:ppn=2
# Request 4 hours of walltime
```

```
# PBS -l walltime=4:00:00
# Request that regular output and terminal output go to the same file
# PBS -j oe
# The following is the body of the script. By default,
# PBS scripts execute in your home directory, not the directory
# from which they were submitted.
# The following line places you in the directory from which the job was submitted.
cd $PBS_O_WORKDIR
# Now we want to run the program "hello". "hello" is in
# the directory that this script is being submitted from,
# $PBS_O_WORKDIR.
echo " "
echo "Job started on `hostname` at `date`"

./exefilename

echo "Job Ended at `date`"
echo " "
```

After editing the above script file, we can use the command qsub to submit it:

qsub script

And then, we can use the command "qstat" to check the job status.

qstat [flags]

We can check a specified job status using a userID to replace "flags" in the line above.

qstat userID

We can delete a job using the command "qdel" from the queue

qdel jobID

where the jobID can be obtained from the command "qstat".

REFERENCES

[1] T. Sterling and J. Salmon, *How to Build a Beowulf: A Guide to the Implementation and Application of PC Clusters*, MIT Press, Cambridge, MA, 1999.

[2] Purdue University, Beowulf Project: http://www.psych.purdue.edu/~beowulf/.

[3] D. Becker, Beowulf Project: http://www.beowulf.org/.

List of Notations

M, A, B: Matrix
$\vec{E}, \vec{H}$:   Time domain vector
$\varepsilon, \sigma, \mu, \sigma_M$:   Scalar
$\tilde{\vec{E}}, \tilde{\vec{H}}$:   Frequency domain vector
$\bar{\bar{s}}, \bar{\bar{\mu}}, \bar{\bar{\varepsilon}}$:   Tensor

Basic parameters and units

Electric intensity:	Volt/meter (V/m)
Magnetic intensity:	Ampere/meter (A/m)
Electric flux density:	Coulomb/square meter (Coul/m^2)
Magnetic flux density:	Weber/square meter (Wb/m^2)
Electric current density:	Ampere/square meter (A/m^2)
Magnetic current density:	Volt/square (V/m^2)
Electric charge density:	Coulomb/cubic meter (Coul/m^3)
Magnetic charge density:	Weber/cubic meter (Wb/m^3)
Permittivity:	Faraday/meter (F/m) ($\varepsilon_0 = 8.854 \times 10^{-12}$ F/m)
Conductivity:	Siemens/meter (S/m)
Permeability:	Henry/meter (H/m) ($\mu_0 = 4\pi \times 10^{-7}$ H/m)
Light speed:	Meter/second (M/s) ($c = 2.998 \times 10^8$ M/s)
Voltage:	Volt (V)
Electric current:	Ampere (A)
Electric charge:	Coulomb (Coul)
Electric flux:	Coulomb (Coul)
Magnetic current:	Volt (V)
Magnetic flux:	Weber (Wb)

About the Authors

Wenhua Yu joined the Department of Electrical Engineering at Pennsylvania State University in 1996. He received a Ph.D. in electrical engineering from the Southwest Jiaotong University in 1994. He worked at the Beijing Institute of Technology as a postdoctoral research associate from February 1995 to August 1996. He has published *Conformal Finite-Difference Time-Domain Maxwell's Equations Solver: Software and User's Guide* (Artech House, 2003) and *Parallel Finite-Difference Time-Domain* (CUC Press of China, 2005, in Chinese) and more than 100 technical papers and two book chapters. He developed two software packages as the primary developer and a group leader, and CFDTD and parallel CFDTD software packages, which have been commercialized in U.S. Navy and U.S. government labs.

Dr. Yu is a senior member of the IEEE. He was invited to be included in *Who's Who in America*, and *Who's Who in Science and Engineering*. He is also a visiting professor and a Ph.D. advisor at the Communication University of China. He founded the Electromagnetic Communication Institute at the Communication University of China and serves as a director.

Dr. Yu's research interests include computational electromagnetics, numerical techniques, parallel computational techniques, and theory and design of parallel computing systems. He has three forthcoming books: *DPS Design and Applications*, *Applications of ACIS and HOOPS to Computational Electromagnetics*, and *Design and Theorem of Low Dissipation Parallel Computing System*.

Raj Mittra is a professor in the Department of Electrical Engineering at Pennsylvania State University. He is also the director of the Electromagnetic Communication Lab, which is affiliated with the Communication and Space Sciences Laboratory of Department of Electrical Engineering. Prior to joining Pennsylvania State University, he was a professor in electrical and computer engineering at the University of Illinois at Urbana–Champaign. He is a life fellow of the IEEE, a past president of the IEEE Antennas and Propagation Society, and he has served as the editor of the *Transactions of the Antennas and Propagation Society*. He won the Guggenheim Fellowship Award in 1965, the IEEE Centennial Medal in 1984, the IEEE Millennium Medal in 2000, and the IEEE/APS Distinguished Achievement Award in 2002. He has been a visiting professor at Oxford University, Oxford, England, and at the Technical University of Denmark, Lyngby, Denmark.

Dr. Mittra is the president of RM Associates, which is a consulting organization that provides services to industrial and governmental organizations, both in the United States and abroad.

Dr. Mittra has published over 800 technical papers and more than 30 books or book chapters on various topics related to electromagnetics, antennas, microwaves, and electronic packaging. He also has three patents on communication antennas to his credit. He has advised over 80 Ph.D. students, about an equal number of M.S. students, and has mentored approximately 50 postdoctoral research associates and visiting scholars in the EMC Laboratories at Illinois and Penn State.

Tao Su is a research associate at RM Associates. He received a B.S., an M.S., and a Ph.D. in electrical engineering from Tsinghua University in 1996, University of Texas at Austin in 1997, and University of Texas at Austin in 2001, respectively. He currently is a member of the IEEE and the IEEE Antennas and Propagation Society, and a reviewer of *IEEE Transactions on Antennas and Propagation, IEEE Microwave and Wireless Components Letters*, and *Antennas and Wireless Propagation Letters*. His research interests include computational electromagnetics, numerical methods, and software development.

Yongjun Liu received a B.S. in electrical engineering from Tsinghua University, China, in 1991, and an M.S. in electrical engineering from the Communication University of China, China, in 1994. From 1994 to 2001, he worked in Beijing Broadcasting Institute as an associate professor. He joined the Electromagnetic Communication Lab at Pennsylvania State University in September 2001. His research interests include computational electromagnetics and its visualization techniques and visual languages.

Xiaoling Yang is a research associate in the Electromagnetic Communication Lab at Pennsylvania State University. He received a B.S. and an M.S in communication and mathematics from Tianjin University in 2001 and 2004, respectively. His research interests include numerical methods, visual languages, and software development.

Index

ADI FDTD 43
Antenna radome 227
Average power 87

Backward difference 2
Base function of FDTD 61
Beowulf PC-clusters 124
BlueGene/L 140

Capacitor simulation 63
Category of MIMD machines 119
Central difference 2
Circular horn antenna 175
Cold plasma 53
Communication mode 129
Complex coordinate PML 27
Contour-path method 35
Courant condition 11
Crossed bent dipole array 185
Crossed dipoles 175

Data compression 113
Data exchange 156, 162, 163
Data exchanging 147
DCT technique 114
Debye material 53
Definition of some concepts 129
Dey-Mittra approximation 35
Diagonal approximation 34
Dielectric conformal technique 40
Diode simulation 66
Directivity 185
Dispersive media 53
Domain decomposition 152
Double negative material 59
DSM system 122

Efficiency of parallel BOR-FDTD 220
Engquist-Majda boundary 19
Equivalence principle 106, 109
Excitation source treatment 170

Far-field collection 167
Flynn's taxonomy 119
Forward difference 2
Frequency response 98

Gaussian derivative signal 72
Gaussian signal 71
GNU 252

High availability 120
High performance computing 120
HUB 245
Huygens' surface 107

IBM BlueGene/L 139
Implementation of MPI library 152
Inductor simulation 64

Linux operation system 246
Local source 73
LOFAR BlueGene/L 140
Lorentz material 53

Maxwell's equations 6
Message-passing interface 229
Mode extraction of time signal 100
Mode extraction 78
MPI data type 241
MPI features 146
MPI functions in C/C++ 236
MPI functions in Fortran 229
MPI library 146

MPI run 131
MPICH architecture 126
MPICH installation and configuration .. 243
MPP system 123
Mur's absorbing boundary 18

Near-to-far-field transform 200
Near-to-far-field transformation 106
Network parameter 103
NFS 251
NIC 245
NIS 249
Nonuniform FDTD method 12
Nonuniform mesh collection 165
Numerical dispersion error 8

Oblique incidence 208
Output power 87

Padé approximation 101
Parallel BOR-FDTD 215
Parallel code structure 130
Parallel efficiency 135
Parallel output 142
Parallel programming techniques 125
Patch antenna array 176, 181
PBS 253, 255
PC cluster 245
PEC boundary 16
PMC boundary 17
PML for ADI-FDTD 47
PML for BOR-FDTD 198
Pole extraction techniques 58
Process status 130

Prony method 101

Reflector antenna 222
Resistor simulation 60
Result collection 166

SBCD FDTD 50
Scattered field formulation 90
Self-consumed power 88
Singularity boundary 207
SMP sytem 121
Staircasing approximation 34
Stretched coordinate PML 26
Subgridding 173
Surface current collection 167
Switch 246

TCP/IP 248
TE mode 83
TEM mode 78
Thin PEC conformal 38
Time convolution PML 27
Time delay 89
TM mode 83
Total power 88
Total/scattered field formulation 94

Unsplit PML 22

Waveguide excitation 171
Window functions 100

Yee's scheme 3

Recent Titles in the Artech House Electromagnetic Analysis Series

Tapan K. Sarkar, Series Editor

Advances in Computational Electrodynamics: The Finite-Difference Time-Domain Method, Allen Taflove, editor

Analysis Methods for Electromagnetic Wave Problems, Volume 2, Eikichi Yamashita, editor

Analytical Modeling in Applied Electromagnetics, Sergei Tretyakov

Applications of Neural Networks in Electromagnetics, Christos Christodoulou and Michael Georgiopoulos

CFDTD: Conformal Finite-Difference Time-Domain Maxwell's Equations Solver, Software and User's Guide, Wenhua Yu and Raj Mittra

The CG-FFT Method: Application of Signal Processing Techniques to Electromagnetics, Manuel F. Cátedra, et al.

Computational Electrodynamics: The Finite-Difference Time-Domain Method, Second Edition, Allen Taflove and Susan C. Hagness

Electromagnetic Waves in Chiral and Bi-Isotropic Media, I. V. Lindell, et al.

Engineering Applications of the Modulated Scatterer Technique, Jean-Charles Bolomey and Fred E. Gardiol

Fast and Efficient Algorithms in Computational Electromagnetics, Weng Cho Chew, et al., editors

Fresnel Zones in Wireless Links, Zone Plate Lenses and Antennas, Hristo D. Hristov

Grid Computing for Electromagnetics, Luciano Tarricone and Alessandra Esposito

Iterative and Self-Adaptive Finite-Elements in Electromagnetic Modeling, Magdalena Salazar-Palma, et al.

Parallel Finite-Difference Time-Domain Method, Wenhua Yu, et al.

Quick Finite Elements for Electromagnetic Waves, Giuseppe Pelosi, Roberto Coccioli, and Stefano Selleri

Understanding Electromagnetic Scattering Using the Moment Method: A Practical Approach, Randy Bancroft

Wavelet Applications in Engineering Electromagnetics, Tapan K. Sarkar, Magdalena Salazar-Palma, and Michael C. Wicks

For further information on these and other Artech House titles, including previously considered out-of-print books now available through our In-Print-Forever® (IPF®) program, contact:

Artech House
685 Canton Street
Norwood, MA 02062
Phone: 781-769-9750
Fax: 781-769-6334
e-mail: artech@artechhouse.com

Artech House
46 Gillingham Street
London SW1V 1AH UK
Phone: +44 (0)20 7596-8750
Fax: +44 (0)20 7630 0166
e-mail: artech-uk@artechhouse.com

Find us on the World Wide Web at:
www.artechhouse.com